TOURISM, DEVELOPMENT AND GROWTH

In a world influenced by an array of new and unpredictable forces, tourism faces challenges to protect and enhance opportunities for future generations of tourists and host communities. Tourism must grow with the strength to respond positively to the changing global environment and societal structure, yet remain compatible with the principles and practice of ecologically sustainable development.

Tourism, Development and Growth presents a comprehensive examination of the concept of sustainability as related to tourism growth. Set within the context of rapid change in contemporary tourism and world trends, the authors examine how new geopolitical, socioeconomic, technological and environmental circumstances are influencing world tourism today.

Distinguishing between sustainable development and sustainable tourism, the authors examine whether and in what form tourism might contribute to sustainable development and growth and the limits to achieving sustainability. Focusing on different types of tourism appropriate to particular situations, the team of leading contributors draw on examples from around the world – Canada, USA, Spain, Belgium, UK, Australia – to explore issues concerning the contribution tourism can make to economic, social, political and environmental advancement of developing countries, and the challenges to and importance of tourism in industrialized nations.

Concern for sustainability, and for the environmental management processes which contribute to it, is of critical importance in the future growth of tourism. If tourism is to continue to grow, that growth must be firmly grounded on the principles of sustainability so that the resources and attraction of the planet remain available for the tourists of tomorrow. In this book the authors identify the opportunities and challenges inherent in achieving sustainable tourism while accommodating growth, and examine new policies and initiatives necessary, from both the private sector and the state, to pursue sustainable tourism growth in an uncertain and changing world.

Salah Wahab is Senior Professor at Alexandria University, Egypt and President of Tourismplan affiliate member World Tourism Organisation. **John J. Pigram** is Professor and Executive Director at the Centre for Water Policy Research, University of New England, Australia, and visiting Professor at the Universities of Waterloo, Canada and Iowa, USA.

D0706402

TOURISM, DEVELOPMENT AND GROWTH

The Challenge of Sustainability

Edited by Salah Wahab and John J. Pigram

London and New York

First published 1997
by Routledge
11 New Fetter Lane, London EC4P 4EE

Simultaneously published in the USA and Canada
by Routledge
29 West 35th Street, New York, NY 10001

© 1997 selection and editorial matter, Salah Wahab and
John J. Pigram; individual chapters, the contributors

Typeset in Garamond by Routledge
Printed and bound in Great Britain by Hartnolls Ltd,
Bodmin, Cornwall

British Library Cataloguing in Publication Data
A catalogue record for this book is available from the British Library

Library of Congress Cataloguing in Publication Data
Tourism, development and growth / edited by Salah Wahab
and John J. Pigram
Includes bibliographical references and index.
1. Tourist trade. 2. Sustainable development. I. Wahab, Salah.
II. Pigram, J. J. J.
G155. AIT59253 1997   97–8840
                                                    CIP

ISBN 0–415–16001–4 (hbk)
ISBN 0–415–16002–2 (pbk)

CONTENTS

Part III Balancing tourism growth with sustainability

Part IV Opportunities and challenges in sustainable tourism

Part V Perspectives on sustainable tourism

ILLUSTRATIONS

FIGURES

TABLES

ILLUSTRATIONS

CONTRIBUTORS

Professor Richard Butler is in the Department of Management at the University of Surrey, Guildford, Surrey, UK.

Professor Chris Cooper is Director of Research at the International Centre for Tourism and Hospitality Research, School of Service Industries, Bournemouth University, UK.

Professor Graham M. S. Dann is in the Department of Tourism and Leisure at the University of Luton, UK.

William C. Gartner is Director of the Tourism Center and Professor of Applied Economics at the University of Minnesota, USA.

Professor Myriam Jansen-Verbeke is in the Department of Social and Economic Geography, Catholic University of Leuven, Belgium.

Craig Marien and *Abraham Pizam* are at the University of Central Florida, Orlando, Florida, USA.

John J. Pigram is in the Department of Geography and Planning at the University of New England, Armidale, New South Wales, Australia.

Professor Richard Prentice is Professor of Tourism and Director of Research in Hospitality and Tourism at Queen Margaret College, Edinburgh, UK.

Stephen L. J. Smith is in the Department of Recreation and Leisure Studies, University of Waterloo, Waterloo, Ontario, Canada.

Professor Norbert Vanhove is Director General of the Westvlaams Ekonomisch Studiebureau, University of Antwerp, Belgium.

Turgut Var is at Texas A and M University, Department of Recreation, Park and Tourism Sciences, Texas, USA.

Professor Boris Vukonic is in the Faculty of Economics at the World Tourism Research and Education Centre, Zagreb, Croatia.

CONTRIBUTORS

Professor Salah Wahab is Senior Professor, University of Alexandria, Egypt and President of Tourismplan, Cairo, Egypt.

Professor Geoffrey Wall is in the Department of Geography, University of Waterloo, Ontario, Canada.

ACKNOWLEDGEMENTS

The idea for this book came originally from Salah Wahab during a meeting in Cairo in 1995 of the International Academy for the Study of Tourism. Salah joined with John Pigram in developing the proposal and approaching potential authors for original contributions around the theme *Tourism, Development and Growth*.

The editors wish to acknowledge the helpful comments of anonymous reviewers and are grateful for the cooperation and support of the contributors and the publishers, Routledge, in particular Matthew Smith, Valerie Rose and Sarah Lloyd.

The valuable assistance of the Centre for Water Policy Research of the University of New England is acknowledged with gratitude. The good efforts of the Director and personnel of Tourismplan are also acknowledged with thanks.

Special thanks must go to Denise Cumming in the preparation of the manuscript. Without Denise's much appreciated help, this book would not have become a reality in such a short time.

Part I

INTRODUCTION

1

THE CHALLENGE OF SUSTAINABLE TOURISM GROWTH

John J. Pigram and Salah Wahab

Concern for sustainability, and for the environmental management processes which contribute to it, is of critical importance in the future growth of tourism. Environmental issues are becoming of increasing significance in the world of tourism, and resolution of these issues will demand far-reaching changes in the way the industry operates. So much so that sustainable tourism has become the focus of widespread attention and research.

This book arose out of the perceived need for a thorough assessment of the parameters of sustainability and its implications in relation to tourism development. Much has already been written in this subject area and the concept of sustainable tourism attracts both critics and advocates. The purpose of this exercise is to harness and bring together a range of views on the theme of sustainability in the context of the growth of tourism as a global phenomenon. The contributions are neither reviews nor case studies, but more, a presentation of provocative points of view which need to be addressed if the ideal of sustainable tourism growth is to be realized.

SUSTAINABLE DEVELOPMENT

The notion of sustainable tourism has its roots in the concept of sustainable development generally, defined by the Brundtland Commission as 'development that meets the needs of the present without compromising the ability of future generations to meet their own needs' (World Commission on Environment and Development 1987: 4). Since then it has been adopted and applied in a wide range of human contexts, in an attempt to address simultaneously both developmental and environmental imperatives.

Although sustainable development has become the catchcry of the 1990s, at the same time the expression has been questioned as a contradiction in terms. This book challenges that point of view, maintaining that conditions can be created so that real development, in terms of human betterment and enhanced life opportunities, is nurtured to be handed on to future generations for their growth and prosperity.

Contributing authors were asked to consider the concept of sustainable development in the context of tourism, addressing the question: can tourism grow in a sustainable manner? Thus an underlying theme in the book is the essential element of sustainability of tourism and whether it is reality or myth. The theme is taken further to explore the limits of sustainability, the relationship between sustainability and the growth of tourism, and the differences, if any, between growth and development.

This is not to say that sustainable tourism has been the experience on the world scene to date. Any number of examples can be quoted of the expansion of tourism in a fashion which is clearly unsustainable. The nature and rate of tourism growth and the environments affected by expanding tourism activity can add to the potential for their own destruction. When degradation of the resource base proceeds to a point where the character and attraction of a tourist destination decline, tourism numbers recede, shortcomings appear in the biophysical infrastructure, and the socioeconomic impact of tourism takes on negative overtones.

The question posed in this book is whether tourism growth must inevitably lead to resource degradation and alienation of participants and host communities. For the tourism sector, the concept of sustainability implies meeting current uses and demands of tourism without impairing the natural and cultural heritage, or opportunities for collective enjoyment of tourists of the future (Hawkes and Williams 1993). Is it possible to foster types and rates of tourism which contribute to enhancement of the host environment, both natural features and in human terms? The thrust of this book is that this can be achieved; tourism growth is inevitable, but it can be managed sustainably and compatibly within environmental constraints.

SUSTAINABLE TOURISM

While sustainability is an integrative concept, most interpretations recognize, in the main, the environmental and socioeconomic dimensions. Compatible human use of environments, and resource management practices that minimize human disturbance of ecosystems and avoid actions with irreversible consequences, are often overlooked.

Moreover, the effect of changes in community views and shifts in attitudes over time towards tourism as an instrument of economic growth and development are almost always disregarded. Again, few interpretations of sustainability clearly set out any spatial context or boundaries. A clear message of this book is that these shortcomings must be offset if sustainability is to have any managerial and developmental relevance to tourism.

The book seeks to contribute to the clarification and systematic appraisal of sustainable tourism and its relationship to socioeconomic growth and environmentally viable resource development. These aims are addressed by way of:

- reference to the problems of delivering sustainable forms of tourism against a background of growth and change in contemporary tourism and world trends
- consideration of the contribution of sustainable tourism to overall growth and development of socioeconomic and environmentally significant resources, including the character and evaluation of tourism destinations
- examination of the task of balancing tourism growth with sustainability in particular contexts
- identification of the opportunities and challenges inherent in achieving sustainable tourism while accommodating growth, and
- examination of the policy initiatives necessary to pursue sustainable tourism growth in an uncertain world.

The book presents a comprehensive and penetrating treatment of the concept of sustainability as related to tourism growth. A positive and coherent thesis is put forward: that tourism can continue to expand globally in keeping with environmental constraints, given a commitment by all stakeholders to ensuring a balance between short-term returns and longer-term viability and sustainability.

STRUCTURE

The book comprises groups of chapters orientated towards the objectives of examining and exemplifying evidence for the growth of tourism in a sustainable manner. Following this introductory chapter, the book opens with a consideration of 'Sustainable tourism in a changing world', and the task of achieving sustainable tourism growth in a world influenced by an array of new and largely unpredictable forces. In such a context, the challenge is to pursue tourism growth in a manner resilient enough to respond positively to a changing global environment and societal structure, yet remaining compatible with the principles and practice of ecologically sustainable development.

The theme of tourism growth with sustainability is explored and then related to the geopolitical, socioeconomic, technological, and environmental changes influencing global tourism. The world is at the threshold of a predictably different age, politically, economically, socially, culturally and environmentally. It is therefore opportune for tourism to be revisited as a significant socioeconomic force and peace-promoting phenomenon. This becomes even more warranted in the light of tourism's elastic and responsive nature and its potential contribution to sustainable development.

The emerging endorsement of growing sections of the tourism industry for more responsible and socially compatible tourism development is recognized, in partnership with community interests and the public sector.

JOHN PIGRAM AND SALAH WAHAB

TRENDS IN TOURISM – LIMITS TO GROWTH?

The theme of sustainable tourism growth is continued in a group of chapters which considers some of the current trends in world tourism, and the questions which arise in regard to possible synergistic relationships between commercial expectations and environmental objectives.

In the second chapter of this section, 'Sustainable tourism – unsustainable development', a distinction is made between sustainable development and sustainable tourism, and the question addressed is whether and in what form tourism might contribute to sustainable development and growth. The focus is not on tourism as an undifferentiated phenomenon, but on types of tourism which are appropriate to particular situations, and the means for assessing and ensuring that they are sustainable.

Sustainable tourism implies a balance between tourism and other existing and potential activities in the interests of multisectoral sustainable development. Despite small steps in the direction of sustainability, the author warns against the endorsement of sustainable tourism by the industry as a marketing gimmick, and questions whether tourism can in fact contribute to sustainable development. Whereas most would agree that it is in the long-term interests of the tourism industry to assure the longevity of the resources on which it depends, relatively little appears to be directed towards maintenance of that natural and cultural heritage.

The following chapter, 'Mass tourism: benefits and costs', begins with definition and documentation of mass tourism and recent trends in participation, both domestic and international. The question is then posed: can these trends be expected to continue? Evidence presented suggests that participation in mass tourism will grow, and the chapter then examines economic impact in terms of benefits and costs.

Economic benefits for a nation, region or local community are described and exemplified, including derived impacts and multiplier effects, income and employment generation, and advantages for less developed countries. Social costs are considered, with a distinction being made between private costs and external diseconomies or incidental costs. Indirect costs, e.g. import substitution, and opportunity costs are also treated.

Overdependence on the growth of tourism is seen as a danger because of its sensitivity and susceptibility to external forces beyond its control. Making tourism sustainable means putting the environment first and encouraging 'new tourism', or the packaging and marketing of non-standardized leisure services, alongside mass tourism.

Concern for sustainability is identified as central to the management of tourism growth in the chapter, 'The contribution of life cycle analysis and strategic planning to sustainable tourism'. The integration of these two mutually compatible approaches to planning provides an organizing framework within which to manage the growth of tourism in a sustainable manner. The adoption of the disciplined, longer-term perspective provided by strategic planning for both destinations and markets ensures that tourism becomes a renewable resource at each stage of its life cycle.

In terms of sustainable tourism, a strategic planning perspective offers an integrated approach to the management of tourist destinations, marked by a common sense of ownership and direction, clear identification of roles and responsibilities, and an agreed set of performance indicators. However, the strategic planning process can be marred by the fragmented, seasonal characteristics of tourism business dominated typically by short-term operating horizons.

When strategic planning is linked with life cycle analysis it is possible to define sustainable options available for destinations according to their competitive position and their stage in the tourism area life cycle. Strategic responses for each stage are identified and the possibilities of new product development are considered in the interests of achieving sustainable tourism outcomes.

Tourism destinations are the focus of the chapter, 'Selective tourism growth: targeted tourism destinations', which explores in some depth, and exemplifies the various meanings of the term 'destination', and the relationship of the evolution of this concept to the objective of sustainable tourism.

Reference is made to particular tourism destinations to demonstrate how clarification of the concept relates to sustainable tourism growth. New planning approaches are needed to conserve the intrinsic qualities of selected tourism destinations. Management and marketing must also be sensitive to particular indigenous characteristics of tourism communities and surrounding areas.

The chapter demonstrates the importance of discussing sustainability of tourism specifically in the context of a particular destination and its inhabitants. Seen through the concept of sustainability, the development of a tourism destination emphasizes better use of space, better terms of presentation, a greater number of more complex tourist activities and services, and the expectation that tourists will enjoy a fulfilling stay.

The section concludes with a chapter entitled 'Modelling tourism development: evolution, growth and decline'. This chapter focuses on the patterns and processes of tourism development with particular reference to the theme of sustainability. A number of general principles of sustainable development are recognized, among them change in the nature, scale and rate of development, the assumption of management to attain long-term goals, and responsibility for avoiding or minimizing impacts on the environment. The challenge to sustainability presented by the emergence of artificial, contrived attractions catering primarily for mass tourism is seen of particular significance.

Understanding the nature of growth and development and explaining its importance in the evolution of tourist destinations are recurring themes in this wide-ranging review. The implications are explored of the author's destination life cycle concept for modelling development of tourism destinations in the context of sustainability. The relationship between the tourism life cycle and carrying capacity suggests that levels of development at each stage should not exceed elements of capacity, however imprecise or measured. To achieve this, some level of control and perhaps regulation are seen as likely to be necessary.

Economies of scale and pressures for development only add to the need for monitoring the manner in which destinations seek to grow in pursuit of marketing advantage. In the words of the contributor, 'The trick is to achieve one of two things: unlimited and indefinite growth, or sustainability. The latter is difficult, and perhaps unattainable; the former is impossible'.

BALANCING TOURISM GROWTH WITH SUSTAINABILITY

The chapters in this section are concerned with process – the task of achieving tourism growth in harmony with sustainability considerations – and in particular spatial and cultural contexts.

'Sustainable tourism in the developing world' is the focus of the opening chapter. The relatively unspoiled nature of many developing countries and their attractive, less sophisticated lifestyle are seen as part of the comparative advantage they enjoy for tourism. Whereas the economic benefits which tourism can bring are undeniable, costs are also likely to be incurred in distortion of value systems, loss of heritage and changes to ways of life.

The chapter traces the contribution which tourism can make to the economic, social, political and even environmental advancement of developing countries. The challenge is to maximize the benefits and minimize the disadvantages, and one way to help the process is through international cooperation and bilateral assistance programmes. The process also requires coherent policy formulation and implementation, and these are typically not well developed in the Third World.

Goals and strategies for effective tourism policies for less developed countries are identified, and the key issues of carrying capacity and community participation are stressed. Tourism in Egypt is used as a case study to illustrate the elements of a successful development strategy for pursuing tourism growth in a sustainable manner.

The chapter 'Challenges to tourism in the industrialized nations' notes the importance of tourism to industrialized countries, both as a domestic industry and an export industry, and its continued growth in the foreseeable future. Some nations such as Australia, Canada and Spain have identified the tourism industry as of strategic importance in their economic future, and have initiated policies to promote the growth of tourism.

Whereas tourism presents challenges to societies, the focus here is on challenges to tourism. Thus the orientation might be described as supply-side in that the chapter examines challenges faced by firms, organizations and the industry at large. Most challenges represent a combination of conceptual, analytical and delivery issues. Each challenge is summarized and the implications briefly described, and the beginnings of a solution to each are outlined.

As tourism growth continues, strategies to deal with these challenges become more urgent, in that they inhibit policy attention from government, limit the ability of tourism businesses to make informed management decisions, and affect the availability and quality of tourism products for consumers. The degree to which tourism can meet these challenges will determine to a significant degree whether it can function as a sustainable industry.

Evaluating a community's sensitivity to tourism development is the first step in planning towards sustainability. The following chapter, 'Implementing sustainable tourism development through citizen participation in the planning process', argues that sustainable tourism cannot be successfully implemented without the direct support and involvement of those who are affected by it. Devising effective means for allowing citizens' involvement in the tourism planning process and encouraging citizens to participate actively in the process is of primary importance for sustainable tourism development.

Various techniques are examined for citizen participation in tourism development, categorized by their effectiveness in achieving certain participatory objectives. Effective participation programmes in tourism development projects will always be a difficult task, and will require a combination of techniques that will work best for a unique set of constituents.

The overview of participatory techniques presented is designed to assist tourism planners and decision makers in achieving sustainable tourism growth. If obstacles to citizen participation can be overcome, tourism outcomes will be enhanced. As pressure for tourism growth gathers strength, the need to promote sustainable tourism development in harmony with community expectations becomes an important element in the planning process.

The formation of touristic images and their relationship to sustainable tourism systems is the focus of the next chapter, 'Image and sustainable tourism systems'. The contrasting meanings of sustainability are discussed with reference, for example, to developers and to host communities, and in the context of growing concern for sustainable development. Since sustainable development, however defined, is a function of use, and destination images can be used to increase use, there appears to be a link between how images are formed and the concept of sustainable tourism systems.

The various ways in which destination images are formed are outlined, and the importance for sustainability of the involvement of host communities in image formation is stressed. In other words, the type of tourism destination image projected to potential tourists has much to do with achieving sustainable outcomes. Determining an appropriate image mix relative to the destination's ability to cope with growth is fundamental.

OPPORTUNITIES AND CHALLENGES IN SUSTAINABLE TOURISM

The chapters in this section are essentially applications or case studies of sustainable tourism and of the possibilities which exist and the impediments to be overcome.

The state of Texas may be identified by many readers with oil wells, cattle ranches and ten-gallon hats. However, Texas is now pursuing economic diversification to counter a downturn in returns from the resources sector, and is exploring alternatives such as tourism to achieve higher employment and income. The chapter, 'Nature tourism development: private property and public use', traces the evolution of nature-based tourism as a tool of economic development in Texas. Nature tourism is seen as a good example of private–public cooperation in maintaining conservation values, while providing for public recreational use of private lands.

A special State Task Force has been appointed to examine the potential of nature tourism in Texas and to recommend opportunities for developing and promoting it. An important element in the mission of the Task Force is to foster sustainable economic growth and development, and environmental conservation through nature tourism. Since 97 per cent of land in the state is privately owned, the key role for the private sector in achieving sustainable tourism growth is obvious. The task is compounded when set against a background of environmental legislation protecting endangered species and habitats.

A number of initiatives are now being undertaken to encourage nature tourism in Texas, while remaining sensitive to the resource base on which success depends. These include tax incentives, legal provisions, education, research and information exchange. In the process, the principle of sustainability should be reflected in the practice of nature tourism and should contribute positively to the wellbeing of local communities and the resource base.

The focus of the following chapter, 'Cultural and landscape tourism: facilitating meaning', is the challenge of facilitating and understanding the meaning of cultural and landscape tourism in the context of sustainability. The effects of tourism on communities and physical environments are far more widely recognized than the effects of tourism on tourists. Equally neglected is how beneficial effects of tourism–environment interaction may be realized through the medium of cultural and landscape imagining. The author seeks explicitly to redress this balance by offering a largely demand-side analysis in which demands for cultural and landscape tourism are reviewed. Particular attention is paid to tourists' environmental preferences and their demands for insight into what is being experienced.

Without knowledge of what tourists are seeking in terms of experiences and benefits of the cultures and landscapes visited, the effects of management strategies are likely to be haphazard and misdirected. Understanding of preferred tourism environments and of segmentation of demand for cultural and landscape tourism is needed if experience-based and benefits-based management of these resources is to be achieved.

A generic framework in the form of a *multiple consumption matrix* is proposed, within which such segmentations may be defined, measured and appraised, and related to tourist motivations, activities and settings. The message is that tourism will only be sustainable if it is simultaneously in harmony with hosts, environment, policy objectives *and* tourists' demands. The multiple consumption matrix offers a means of capturing these linkages and applying them to the sustainable development of tourism destinations.

The chapter 'Urban tourism: opportunities and networks' addresses the options for developing tourist products based on the cultural resources of cities. The objective is to contribute to the understanding of cultural tourism, the opportunities for urban tourism growth and the need for anticipatory policies and planning models for the sustainable management of resources and visitors. Cultural tourism is seen to play a key role in the development of new forms of tourism, characterized by flexibility, segmentation and integration between products, in contrast to the standardized and packaged mass tourism market.

A case study of the Belgian city of Bruges demonstrates the many alternative approaches to management of urban resources and of tourists in accordance with the principles of sustainability. Using cultural tourism as a stimulus for urban growth needs to be matched by integrated planning to enrich the tourist experience and safeguard resources for future generations. The Bruges case study is an excellent example of the multidimensional character of sustainable tourism and the way in which positive planning initiatives can help accommodate the ideology of sustainability in a market-driven growth model of tourism.

The final chapter in this section, 'The green grass of home: nature and nurture in rural England', makes the point that tourism is not just about resources and biophysical phenomena, but also about people and culture and heritage, and how the latter are variously combined to yield touristic experiences.

The discussion is set in rural England and numerous examples are presented in raising questions about the sustainability of past-fixated tourism. Fascination with nostalgia, it is argued, is derived in part from childhood associations and a sense of greater security. However, subjective feelings about the past may have little to do with actuality. Rather,

11

Olde England is being promoted to the point where the process of mass tourism may threaten to degrade the features which make it so attractive.

The author poses the question: how enduring and environmentally friendly are forms of tourism which depend for survival on marketing contrived and fanciful representations of yesteryear? The risk of failure and environmental stress is greater where the marketing is imposed from outside with little or no input from local communities. Authenticity of experience often resides more in people than in a given place, but there must be a balance between host privacy and visitor satisfaction. The concern is that the proliferation of nostalgia-laden images of the country-side leading to touristic activity may destroy the very locales which they seek to promote.

PERSPECTIVES ON SUSTAINABLE TOURISM

The concluding chapter, 'Tourism and sustainability: policy considerations', considers again the concept of sustainability and asks whether sustainable tourism can be a reality. The challenge for tourism is to protect and enhance opportunities for future generations of tourists and host communities in a time of change. Thus the chapter echoes some of the concerns put forward in the opening chapter regarding the additional impediments on sustainability as tourism expands and adapts to new geopolitical, technological, socioeconomic and environmental circumstances.

A range of viewpoints is presented on sustainable tourism growth, with carrying capacity in all its facets as a focal point. Basic guidelines for achieving sustainable tourism are put forward and the duality of growth and development is stressed. Development implies interference or manipulation in the process of growth, in the interests of sustainability of tourism expansion. Key objectives are set down for achievement of this in accordance with Agenda 21.

The chapter concludes with a discussion of policy considerations for tourism sustainability and growth and the respective roles of private sector and state. A policy checklist is included to guide progress towards balanced and sustainable tourism growth.

SUMMARY

The 1992 Earth Summit gave rise to a worldwide increase in awareness of the links between ecologically sustainable development and environmental management. In the world of tourism this has been translated into growing endorsement of sustainability as an essential element in the development and operation of tourist facilities. Whereas there remains much scepticism as to the extent of the commitment to sustainable tourism in a period of unprecedented growth, the views presented in this book are generally positive and encouraging. Clearly, however, if tourism is to continue to grow, that growth must be firmly grounded in the principles of sustainability, so that the resources and attractions of the planet remain available for the tourists of tomorrow.

REFERENCES

Hawkes, S. and Williams, P. (1993) *The Greening of Tourism,* Burnaby BC: Simon Fraser University, Centre for Tourism and Policy Research.

World Commission on Environment and Development (1987) *Our Common Future,* Oxford: Oxford University Press.

Part II

TRENDS IN TOURISM – LIMITS TO GROWTH?

2

SUSTAINABLE TOURISM IN A CHANGING WORLD

John J. Pigram and Salah Wahab

Tourism in the twenty-first century must anticipate a future marked by changing global relationships and societal structures, technological innovations, and growing spatial awareness and environmental concern. Major shifts in the nature and scale of tourism are already in evidence as the industry struggles to respond to a range of forces at work. Change and the change agents involved are powerful and positive forces if harnessed constructively. Rather than opposing change, or merely accepting and accommodating change, the tourism industry must manage change to its advantage and that of the environment which nurtures it. Endorsement and application of the concept of sustainability and of best-practice environmental management offer compelling evidence of how change can be harnessed to contribute towards the achievement of environmental excellence. Although tourism flourishes best in conditions of peace, prosperity, freedom and security, disturbance to these conditions is to be expected. The industry response must be sufficiently resilient to generate opportunities for the growth of tourism in keeping with the dynamics of a changing world and increasing concern for ecologically sustainable development.

INTRODUCTION

Since the Earth Summit held in Rio de Janeiro in 1992, pressure has grown for the tourism industry to lift its environmental performance in common with other economic sectors, and to work towards ecologically sustainable forms of tourism development. At the same time, planners and developers seeking to create satisfying settings for tourists into the next century must anticipate a world influenced by a new and bewildering array of forces, many of which are difficult to predict. Given such uncertainty, the challenge is to pursue tourism growth in a manner resilient enough to respond positively to a changing global environment and societal structure, while remaining compatible with the principles and practice of ecologically sustainable development.

SUSTAINABILITY CONCEPTS

The era of environmental concern ushered in by the World Conservation Strategy and the Brundtland Commission in the 1980s was given renewed impetus following the Rio Summit and the adoption of Agenda 21. Most governments around the world have now committed themselves to ecologically sustainable development in a wide range of policy areas.

The concept of sustainability expresses the idea that humankind must live within the capacity of the environment to support it (Jacobs 1995). Sustainable development is that which 'meets the needs of the present without compromising the ability of future generations to meet their own needs' (World Commission on Environment and Development 1987: 43). It is difficult to argue with such a desirable, if somewhat fuzzy approach. However, it becomes more meaningful when supplemented by more specific objectives and guiding principles. These have been articulated by any number of national and international agencies and organizations, and typically embrace components of:

- conservation and enhancement of ecological processes
- protection of biological diversity
- equity within and between generations
- integration of environmental, social and economic considerations.

If this all sounds familiar, it is not surprising. Indeed, the concept of ecologically sustainable development has been challenged as merely a reworking of longstanding philosophies about conservation and stewardship of resources for the future. What is new is the widespread endorsement of the concept and the incorporation of a proactive environmental dimension into corporate planning and management (Murphy 1994). Sustainability implies a fresh approach to planning and a renewed commitment to use resources within the capacity of the environment to sustain such use.

Support for ecologically sustainable development is now emerging strongly in the tourism sector, as the logical way of balancing environmental concern with growth and development of the industry.

TOURISM AND SUSTAINABILITY

Tourism is reputed to be the world's largest economic sector (World Tourism Organisation 1995), and one with a vested interest in the environment and the resource base on which it depends. The environment, considered here in the broad sense as encompassing socioeconomic and cultural phenomena as well as biophysical elements, represents both a resource and an opportunity for tourism, as well as a potential constraint on the manner of its development.

Ideal tourism settings grow out of complementary natural features and compatible social processes. More often than not, however, tourism in today's world finds it necessary to manipulate and modify the environment to suit its purpose. The consequences are not always predictable. Tourism certainly can contribute to environmental degradation and be self-destructive; it also has potential to bring about significant enhancement of the environment (Pigram 1991). With tourism-induced change, an important issue, for example, is irreversibility, which in part is a function of the resilience of the resource base, along with the spatial and temporal pattern of impacts, and the scope for compensatory managerial response.

An influential conference held in Canada on 'Global opportunities for business and the environment' concluded that sustainable development holds considerable promise as a vehicle for addressing the problems of modern tourism (Tourism Canada 1990). Clearly, what seems to be occurring is recognition of the interdependence between environmental and economic issues and policies, and ultimately, acceptance of the notion that sound environmental management in tourism does not merely cost, it pays. Of course, convincing decision makers of the good sense, both environmentally and commercially, of the merits of a sustainable approach to tourism may not be easy. It can be difficult to demonstrate conclusively the costs over time of environmental degradation, especially when set against the more immediate returns forgone by adopting a more restrained pattern of development. However, some significant initiatives are being pursued, and several of these are detailed in a 'Casebook of best environmental practice in tourism' (Hawkes and Williams 1993). The tourism industry of the 1990s is showing commendable preparedness to apply the principles of best practice environmental management to its activities, especially at the larger-scale, corporate level. This point is taken up later in this chapter, along with reference to the important role for governments and the public sector in furthering the process.

TOURISM AND CHANGE

If the process of achieving sustainable forms of development in a growing tourism industry is a formidable challenge, the task takes on additional dimensions when set against the many forces for change facing the industry as it moves into the next century (Theobald 1994). New spatial forms and settings must be anticipated as tourist developers seek to service a diverse and expanding array of interests and opportunities. Tourism can be a very volatile industry, sensitive to changes in perception and taste, and to altered biophysical, economic or political circumstances (Long 1993).

Thus the theme of change is constant on the tourism scene. In a business sense, the professional component of tourism is constrained to keep abreast of change, and be aware of, and respond to (if not articulate) changes in travel preferences, in technology, and in the many factors which help fashion

19

the travel market. A danger is that in these circumstances, sustainability may only be pursued in a financial or business management context. Ecologically sustainable tourism development may be seen as too expensive and irrelevant to business success.

Much research interest focuses on the responses of tourists and tourism to changing socioeconomic circumstances, to political, cultural and attitudinal changes, to improvements in spatial awareness and communication, and to growing environmental concern for sustainability. The potential implications for tourism of these influences, in particular, the way they appear to have intensified in the 1990s, have been stressed by Ritchie (1993):

> the dramatic political changes that have occurred in recent years are only manifestations of more deepseated social and cultural transformations which reflect changes in human priorities concerning the way the populations of the world wish to live. Tourism, as a phenomenon, is clearly affected as much and perhaps more by these changes as any other sector.
>
> (Ritchie 1993: 201)

Tourism certainly is a highly sensitive and vulnerable activity and it is not without reason that tourists have been described as 'shy birds', who can be scared off by any number of real or perceived threats to safety, health and property, or financial wellbeing. Tourism shares with recreation generally, attributes of voluntary, discretionary behaviour. People are free to choose to become tourists and to decide location, timing, duration, mode of travel, activities and costs to be incurred. Any one of these attributes may be modified or dispensed with by unforeseen or uncontrollable factors. Moreover, the process of choice is imperceptibly influenced by pervasive adjustments to lifestyle, social mores, traditions and culture. The more environmentally aware tourist of the future may well opt for, and even demand, evidence of a commitment to sustainable forms of tourism.

Changes impinging upon the nature and scale of tourism in the modern world may be categorized as geopolitical, socioeconomic, technological and environmental. Some comments follow on each of these categories and how their influence may lead to fluctuations in the patterns and characteristics of world tourism. In a world marked by multiplicity of change agents, the challenge of achieving tourism growth in a sustainable manner becomes an even greater concern.

GLOBAL GEOPOLITICS

Some of the most dramatic events to impact on tourism have occurred in the political geography of the planet. The 1990s have seen great changes in Europe, the Middle East, Africa and Asia. Strong efforts will be necessary to ensure that the opportunities for tourism created by these changes are not ephemeral, but long-term and sustainable.

The freeing-up of access to the countries of Central and Eastern Europe following the collapse of the Soviet Union, the removal of the Berlin Wall and the reunification of Germany is a case in point. Moreover, the process of change is multifaceted and ongoing. For example, internecine conflict in Chechnya and other enclaves continues to threaten what internal stability remains in the former Soviet Union. At the same time, evolving political and economic philosophies are altering the character and cohesiveness of Russian society. Meanwhile, civil strife following the breakup of Yugoslavia continues to overshadow tourism in the Balkans and along the Adriatic Coast.

Outcomes for this region as a tourist destination must remain speculative, with the balance tending towards a more positive and sustained trend in tourism flows in the future. In the process, opportunistic exploitation of speculative tourism ventures should be avoided in the interests of longer-term programmes of sustainable development.

Further west, the European Community is pushing steadily towards a common currency and relaxation of controls on visitors between member states. Again, these moves should provide a boost to tourism, which should become more pronounced following the opening of the Channel Tunnel.

The changed geopolitical situation in the Middle East offers some of the most convincing evidence of how tourism can react to perceived threats to security and safety and yet subsequently recover. Beirut and Lebanon, now relatively free from hostilities, are re-emerging as prime tourist destinations in the Eastern Mediterranean. Fine hotels, once pock-marked or reduced to rubble, are being rebuilt as foreign investment sees an opportunity to restore this area to its former importance for tourism.

Similarly, the prospects of peace between Israel and neighbouring states present further opportunities to boost tourism in the region. As the horror of the Gulf War and repression in Iraq and Iran fade, these nations too should share in this resurgence. Situations can change rapidly but the peace momentum promises new freedom of travel within the Middle East, and with it an upsurge in tourist flows. This new beginning brings with it not only the promise of financial recovery, but also the opportunity to develop the region's tourist potential in a more sustainable manner.

One of the most striking examples of geopolitical influence on a country's fluctuating tourism fortunes is Egypt. The much publicized attacks on tourists by militant groups of Islamic fundamentalists led to a catastrophic drop in numbers of western tourists in the 1990s. Even visitors from Britain, traditionally the largest group from the West, showed a 17 per cent decline in arrivals in Egypt in 1994. It is estimated that the country has lost some US$2 billion in tourist earnings since the attacks began, although these were largely confined to Upper Egypt (Binyon 1994). Whereas tourist reaction to safety concerns is a matter of perception, the threat should not be overstated. It should be emphasized that since 1992, over seven million tourists have visited Egypt and only seven have been killed in terrorist attacks (El Beltagui 1995).

Much of the loss to Egyptian tourism has been at the lower end of the market in package tours. One of the most important groups financially, Arab tourists from the Gulf States, has scarcely been affected. Egyptian tourism also stands to gain further from the 'outbreak of peace' in the Middle East. With the opening of international borders, tourism is now being promoted in the 'Red Sea Riviera' embracing Sinai (Egypt), Eilat (Israel) and Aqaba (Jordan) in close proximity. Egypt is making strenuous efforts to counter recent setbacks and broaden its appeal to tourists beyond the archaeological heritage of the Nile Valley, while at the same time pursuing the goal of sustainability in a demanding environment (Wahab 1995).

Sub-Saharan Africa too has seen far-reaching changes to political circumstances which are reflected in fluctuations in tourism activity. Rwanda, once known as the 'Switzerland of Africa', has been effectively removed from the tourist scene by civil strife, while instability in several other nation states is hardly conducive to attracting tourists. In many instances fears over personal safety are heightened by health risks and exposure to disease. On the positive side, Zimbabwe, Botswana and Namibia are being rediscovered as destinations for ecotourism. Further south, the ending of apartheid and the emergence of majority rule in South Africa promise a new era for tourism in that region, and an opportunity to learn from the mistakes and successes elsewhere in implementing sustainable development.

South and East Asia complete the picture of significant restructuring of the tourism scene in a large part of the globe. Geopolitical change is again in evidence. The opening-up of China to the world is reflected in impressive flows of tourists. From the openly commercial ports of Hong Kong and south-east China to the Silk Road of the interior, China is promoting new tourism destinations and experiences. For the period 1985–92, China recorded the fastest annual growth in arrivals worldwide of almost 13 per cent (World Tourism Organisation 1993). Outbound tourism from China must await further relaxation of travel restrictions, and the cessation of British rule in Hong Kong in 1997, which may act as a catalyst in the transition (Ap 1995). Reports suggest, however, that environmental concerns may come second to economic gains in the expansion of tourism.

More generally, the countries of South-East Asia are generating substantially increased tourist flows. In 1993, East Asia and the Pacific was the fastest growing tourist region, with Australia emerging as the destination of choice. Australia expects to see more than two million visitors from South-East Asian countries by the year 2000, with another half-million from Japan. South Koreans have rapidly become one of the leading tourist groups in the Pacific, along with Taiwan and Singapore. The remarkable turnaround in attitudes towards Vietnam is also symptomatic of the changes occurring in the region. Whereas some twenty years ago the only visitors to Vietnam were military, the country is now actively encouraging and catering for an influx of tourists.

Global geopolitics can work negatively and positively for the development of tourism. Changed geopolitical circumstances can present an opportunity to redirect the growth of tourism down the path towards sustainability, while at the same time contributing to reconstruction of devastated economies, shattered societies and degraded environments.

SOCIOECONOMIC FLUCTUATIONS

It is not necessary to go further than the most recent global recession to trace the impact which an economic downturn can have on tourism. Tourist flows worldwide experienced severe decline in the early 1990s as discretionary incomes fell and unemployment rose. Economic recovery has been slow and patchy, matched by a positive, if hesitant, response in tourist activity, and regional contrasts in a return to previous levels (World Tourism Organisation 1995).

The shock of economic recession and subsequent restructuring of national economies helped focus renewed attention on the longer-term potential of tourism for boosting economic activity and employment. Butler (1995) remarks that reductions in public sector expenditure which have accompanied widespread economic restructuring, have prompted governments to encourage private sector investment in tourism projects, either separately or as joint ventures. Construction and operation of casinos in several Australian states are examples.

Societal changes normally do not exhibit the same cyclical characteristics as those of an economic nature. Rather, they are evolutionary and cumulative, and sometimes almost imperceptible. Ageing of western populations has been accompanied by a measure of affluence, increased unobligated time, and the desire to remain active, among older age groups. It is no longer unusual to find 'elderly' people undertaking strenuous (and expensive) tours, uninhibited by the misperceptions, restrictions and taboos of a time past.

At the other end of the demographic spectrum it is becoming commonplace for younger, unaccompanied people to enter the tourist flows. Youth tourism, stimulated by a range of motives and underpinned again by 'free' time, energy, a spirit of adventure, and perhaps restrained expectations, is a growing segment of the tourism market. Changes in attitudes to marriage and family, to gender, sex and racial discrimination, are additional forces helping to fashion the tourism of the 1990s.

Whereas socioeconomic fluctuations can create difficulties for the tourism sector, those same societal changes and economic adjustments can lead to a more discerning tourism clientele. A greater role for communities in tourism developments and more searching scrutiny of tourism proposals should help achieve the objective of sustainability.

TECHNOLOGICAL INNOVATIONS

Technology is a powerful influence on modern tourism. Technological break-throughs can change the sustainability status of a tourism business or destination overnight, but they can also offer effective means towards long-term sustainability in an ecological sense.

Tourism, as with many other sectors of the economy, has experienced the shock of technological change. Improvements in transport and communications have decreased the friction of distance and made the greater part of the globe accessible. The advent of long-distance, large-capacity aircraft has made mass tourism a reality at the international scale, and with it the capacity to make or break intermediate points on air routes. Airlines crossing the Pacific now routinely overfly Nadi Airport in Fiji, with detrimental effects on the economy of the Islands. Honolulu is frequently bypassed by long-range aircraft flying directly point-to-point. Even on the North American mainland, an airline's choice of hub can have repercussions for regional economies based in part on servicing aircraft, equipment, crews and passengers.

High speed, computer-based communication and reservation facilities are now an integral part of the global tourism network. Not only do these facilities enable instantaneous links across the world, they also have added immeasurably to levels of awareness, both of tourists and those serving the travelling public. With awareness and exposure can come stimulated demand for hitherto little known destinations. This, coupled with the ability to move vast numbers of people great distances in relatively short periods of time, means that few parts of the planet can any longer be regarded as out of the reach of tourists. This is a somewhat disturbing prospect, given the shrinking availability of sites to cater for the growing interest and involvement in ecotourism, nature-based tourism and unspoiled destinations (Brackenbury 1993). The natural environment is a growing magnet for tourists and great care will be needed in managing access to and use of natural sites so that technological advances do not threaten ecological sustainability.

Perhaps the most exciting and powerful tool of technology has yet to impact fully on the tourism scene by way of the Internet's World Wide Web. The potential of the rapidly expanding Internet to create and disseminate knowledge relevant to tourism marketing is receiving increasing attention (Hawkins 1995). Accessed through the Internet, interactive media can create new sales opportunities for tourism enterprises and services. Hawkins stresses the importance for the tourism industry of understanding how to use the Internet, the online services, and interactive electronic media, to market and sell travel products. Already, travel companies have established itinerary planning systems, and tour operators, travel suppliers and destination management groups have pages on the World Wide Web. It is estimated that more than 100 million people will be connected to the Internet by the year 2000, with instant access to relevant, in-depth, up-to-date information about any country or destination. Used interactively, the Internet offers the facility to select tourism sites and activities based on complete product information, and the promise of a fulfilling tourism experience, yet in a sustainable environmental context.

As with many of the technological innovations being introduced into the tourism industry, reliance on the 'information super-highway' needs qualification. To some, the technology is seen as dehumanizing, with 'online' images poor substitutes for reality and authentic experiences. Hawkins (1995) points also to user confusion and dissatisfaction, as well as the lack of trained specialists, cost, and questions of security on the Internet, as issues to be addressed in harnessing these technological tools for the benefit of tourists and an ecologically sustainable tourism industry.

ENVIRONMENTAL CHANGE

One of the most fundamental challenges facing tourism is adjusting to changes in the environment, both biophysical and human. Change in nature is the norm, but natural changes are typically slow and amenable to human adjustment. Changes brought about by human intervention, no matter how well intentioned, are often sharp and harsh, and likely to evoke a negative response. The fact that a site has been selected as a destination for tourists will inevitably lead to changes in the ambient environment. The features which attracted tourists in the first place can lose their appeal with intensification of use. Alien structures, introduced to the environment ostensibly to support or protect the resource base, do not always sit comfortably with the natural environmental structures, nor will they necessarily have a positive impact in attracting or retaining tourists. This is the very antithesis of sustainable development.

Typically, the initial force motivating tourists is the landscape, encompassing attributes of both the physical and social environment. As a tourist destination matures, these attributes undergo change and the landscape of tourism reflects the imprint of increasing numbers of visitors. The transformation of the Palm Springs area of California from a desert spa and vacation site to 'the golf capital of the world' is a case in point (Pigram 1993). Created attractions are added, facilities are provided, and infrastructure is expanded to present a new blend of structures, activities and functions to service a different clientele. Not all such amendments are ill advised, and change does not necessarily equate with degradation. If carried out sensitively, tourist development can contribute to substantial upgrading of the environment and enhance visitor enjoyment and ecological sustainability. At the same time, the makeup, if not the volume, of tourist flows is likely to alter as the destination takes on a different character. However, it could well be that there would be as many tourists attracted as are repelled by the metamorphosis. Palm Springs and surrounding resorts, for example, now attract over two million visitors each year (Pigram 1993).

Several observers have alluded to what they see as a cycle of tourist area development (Butler 1980; Cooper 1990). In an earlier paper, Plog (1972) expressed concern about what he appeared to see as the inevitable sequence of decline of resort areas:

> Destination areas carry with them potential seeds of their own destruction, as they allow themselves to become more commercialized and lose their qualities which originally attracted tourists.
>
> (Plog 1972: 4)

However, it is important to note that decline is not inevitable, and with sound planning and management it is possible for the downturn to be checked, rejuvenation achieved, and viability sustained. Advances made in water management in the arid region of Palm Springs are part of the success story of sustainable tourism in a desert setting (Pigram 1993).

There appear to be several modes of expression of tourism–environment interaction and the simplistic view of resisting change and isolating the environment in order to protect it *from* tourism may not be the preferred option. A more positive alternative could be to sustain and enhance the environment *for* tourism. According to Gunn (1972), tourism and protection of the resource base are more alike than contradictory. Gunn maintains that the demands of tourism, rather than conflicting with conservation, actually require it. Otherwise the very appeal which attracts visitors to a destination will be eroded and with reduced satisfaction will go any chance of sustained viability. Ideally, tourism development and ecological sustainability should be complementary and mutually reinforcing.

INDUSTRY RESPONSE

The forces categorized above – geopolitical, socioeconomic, technological and environmental – are already highly visible and instrumental in fashioning the turbulent world of today's tourism. These same forces are also exercising the minds of decision makers charged with anticipating the future face of tourism.

At the tenth General Assembly of the World Tourism Organisation, in Bali in October 1993, a round table was held on 'Trends and challenges in tourism – beyond the year 2000'. Discussions were summarized under a number of headings, e.g. growth, technology, industry structures, social impacts, role of government, etc. (Plimmer 1993) and have much in common with those identified earlier in this chapter.

This reflects strong endorsement by the tourism sector of the drive in industry generally for more responsible and socially compatible forms of development. Environmental issues and social impacts are prompting a growing level of community concern over the types of tourism to be tolerated. This will create the need for new alliances, greater consultation, and even power sharing, in forging new tourism products to fit community values and environmental constraints.

Industry representatives at the round table addressed a number of concerns regarding the future shape of world tourism, especially in regard to emerging tourism markets and destinations, and the increasing importance given to environmental issues, environmentally compatible tourism developments and nature-based tourism.

A representative of Inter-Continental Hotels echoed these concerns for the global hotel industry (Collier 1993). However, as with the issues raised earlier in this chapter, they were identified as opportunities and challenges, not necessarily problems. Among the concerns were:

Technology: Adapting to and exploiting technology more quickly especially in routine tasks to give an ever more expensive employee more time face to face with an ever more difficult to get customer.
Environment: Making sure the new hotels we build are compatible with the cultural, social and economic environment within which we operate.

(Collier 1993: 25)

Clearly, there is recognition on the part of various components of the tourism industry that the future has already arrived, and that they must respond to a range of strategic signals pointing towards a tourism scene of markedly different dimensions. Among key indicators, the following stand out:

- Socioeconomic and demographic change
- A revolution in information/communication technology
- More knowledgeable and demanding customers
- Deregulation of the market and emergence of mega-operators
- Public sector–private sector alliances
- Need for human resources training
- Community pressure for sustainable forms of tourism.

A key element of tourism success in the future will be the ability and willingness of the industry to recognize change and use it to its advantage for longer-term sustainable growth. More and more, community and public sector support for tourism development is likely to be conditional on the industry entering into a partnership to pursue ecologically sustainable forms of tourism.

COPING WITH CHANGE

The changes and change agents reviewed in this chapter should not be perceived in negative terms, or as threats to well established modes of practice. Indeed, tourism, perhaps more than most other human activities, should welcome change and take up the opportunities that change offers. Change is a powerful and positive force which, when harnessed constructively, challenges individuals, groups and organizations to perform to their optimum capability.

> Encouraging people not to dwell on change itself, but to accept it as natural and inevitable, is a key part of creating the willingness and mindset to not only cope with, but to thrive on the challenge.
>
> (Strong 1995: 5)

The view that change represents an opportunity and not necessarily a problem for tourism underpins the strengthening moves in many parts of the tourism industry to manage change in a positive manner towards environmental enhancement and sustainability.

> Enough is now known to require any tourist development, new or existing, to abide by the principles of sustainability. . . . Industry is thoroughly receptive to sensible controls on development and properly considered management schemes to protect and enhance tourist destinations. . . . The end of the environment is the end of tourism.
>
> (Brackenbury 1993: 17)

These statements find support in the growing evidence that tourism undertakings, at least at the larger-scale, corporate level, are prepared to foster environmental excellence through the endorsement of what has become known as best practice environmental management (Pigram 1995). Best practice is now widely accepted and promoted in manufacturing industry as the essential means of managing and organizing business operations and achieving quality assurance. When linked to quality assurance, the broader concept of best practice environmental management is of direct relevance to tourism in a competitive and changing world.

Best practice environmental management calls for radically different organizational structures and attitudes to bring about continuous improvement in a firm's environmental performance, in response to changing physical and economic circumstances, market forces, technological breakthroughs and government policies. Translated to the tourism industry, best practice environmental management becomes a vehicle for facilitating and harnessing change. Environmental excellence is fostered through enlightened management practices incorporating new, cleaner technologies, and an emphasis on resource conservation, recycling, recovery and reuse, and attitudinal change, in pursuing progress towards sustainability.

On the international scene a number of large hotel corporations, for example Canadian Pacific, Inter-Continental, Ramada, etc., have implemented a range of effective environmental measures (Checkley 1992; International Hotels Environmental Initiative 1992; Hawkes and Williams 1993). Environmental codes of ethics for tourism operations are also being introduced (Australian Tourism Industry Association 1990).

The expectation that tourism establishments will be developed and managed at the highest standards of environmental excellence needs to be tempered by knowledge that significant impediments stand in the way of rapid and widespread enhancement of environmental standards. The expertise, expense and long-term commitment of resources involved in lifting environmental performance inevitably means that the adoption of best practice environmental management is 'currently, a minority activity, confined, in the main, to a few large firms' (Goodall 1995: 34). The challenge is to raise concern for the environment among the smaller and more numerous establishments, and achieve something of a 'trickledown' effect in the spread of environmental best practice to all levels of tourism activity.

As support and enthusiasm for 'greener' tourism, ecotourism and nature-based tourism gathers momentum, pressure on operators at all levels to lift their 'environmental game' could well be reinforced by market forces. It is not inconceivable, in a more environmentally aware world, that visitor preference might be directed towards those establishments which can demonstrate a superior environmental track record. In a travel market which continues to gather strength, 'green tourism' should be seen not only to offer new experiences and opportunities, but to make economic good sense (Dingle 1995).

CONCLUSION

New forms of tourism bring with them new environmental challenges and more demanding standards for sustainable development. With ongoing expansion the tourism sector can expect to face increasingly stringent conditions on growth and development and be called upon to justify its claims on environmental resources with a firm commitment to their sustainable management. Rather than opposing change, or merely accepting and accommodating change, the tourism industry must take the high ground and help orchestrate and manage change to its advantage and that of the environment which nurtures it.

Environmental considerations are not the only changes facing the uncertain world of tourism, although they may ultimately prove to be the most persistent and demanding through time. Tourism also needs conditions of peace, prosperity, freedom, security, and the absence of threats to health and wellbeing, in order to flourish. Dramatic events can alter the political geography of the globe; natural and human-induced disasters can destroy the attraction of established tourist destinations; severe fluctuations in economic circumstances can affect profoundly propensities to travel, as well as the viability of tourism business and the prospect of public sector support; technology in all its facets can alter irrevocably the spatial and temporal dimensions of tourism operations.

Added to these uncertainties is the problem of long-term planning horizons for investment in tourism versus short-term decision making of tourists indulging in unpredictable and seemingly fickle choice behaviour. The developer investing in tourism is constrained by asset fixity, characterized by site-specific property and capital investments which frequently are also functionally specified. It is not always easy to change a 1,000-room, five-star hotel into a hospital or retirement home when tourism patronage declines. In the world of tomorrow tourism will need to draw on a selection of tools in its drive for economic and ecological sustainability. Advanced technologies, changed managerial behaviour, new environmental laws, better planning and development control procedures, and innovative environmental management systems will be critical means towards achieving and maintaining sustainable tourism while accommodating growth (Buckley 1995).

In such circumstances, anticipating or predicting the nature and pace of change is a difficult and potentially hazardous task. Forecasts of possible tourism scenarios range from the fanciful prophecies of science fiction to more considered statistical extrapolations of current trends. One thing is certain: the tourist map of the twenty-first century will be different from that of today. As in many phases of life, success and fulfilment will go to those components of the tourism industry which can learn from the past, adapt to the present, and plan to manage the challenges and opportunities of the future to their advantage, that of the tourist, and of the environment on which tourism depends. Sustainable growth of tourism in a changing world is attainable given the commitment of all stakeholders – tourists, the community and the public and private sector.

REFERENCES

Ap, J. (1995) 'Hong Kong tourism: 1997 and beyond', Conference of the Asia Pacific Tourism Association, Pusan, September.

Australian Tourism Industry Association (1990) *Environmental Guidelines for Tourism Developments*, Canberra: ATIA.

Binyon, M. (1994) 'Middle East woos tourists', *The Times*, 24 November, 16.

Brackenbury, M. (1993) 'Trends and challenges beyond the year 2000', *WTO Round Table on Beyond the Year 2000: Tourism Trends and Challenges*, Madrid: WTO, 14–20.

Buckley, R. (1996) 'Environmental audit', in F. Vanclay and D. Bronstein (eds) *Environmental Impact Assessment: State of the Art*, New York: Wiley.

Butler, R. (1980) 'The concept of a tourist area cycle of evolution: implications for management of resources', *Canadian Geographer* 14: 5–12.

——(1995) 'Introduction', in R. Butler and D. Pearce (eds) *Change in Tourism*, London: Routledge, 1–11.

Checkley, A. (1992) 'Accommodating the environment: the greening of Canada's largest hotel company', *Proceedings of ISEP Conference on Strategies for Reducing the Environmental Impact of Tourism*, Vienna: International Society for Environmental Protection, 178–89.

Collier, R. (1993) 'Beyond 2000: trends and challenges in the global hotel industry', *WTO Round Table on Beyond the Year 2000: Tourism Trends and Challenges*, Madrid: WTO, 21–6.

Cooper, C. (1990) 'The life cycle concept and tourism', Conference on Tourism Research into the 1990s, University of Durham.

Dingle, P. (1995) 'Practical green business', *Insights*, English Tourist Board, London, C35–45.

El Beltagui, M. (1995) Opening Address, 4th Biennial Meeting, International Academy for the Study of Tourism, Cairo, June.

Goodall, B. (1995) 'Environmental auditing: a tool for assessing the environmental performance of tourism firms', *The Geographical Journal*, 161 (1): 29–37.

Gunn, C. (1972) *Vacationscape: Designing Tourist Regions*, Austin: Bureau of Business Research, University of Texas.

Hawkes, S. and Williams, P. (1993) *The Greening of Tourism*, Burnaby BC: Simon Fraser University, Centre for Tourism and Policy Research.

Hawkins, D. (1995) 'Travel marketing on the Internet', 4th Biennial Meeting, International Academy for the Study of Tourism, Cairo, June.

International Hotels Environmental Initiative (1992) *Environmental Management for Hotels: The Industry Guide to Best Practice*, London: IHEI.

Jacobs, M. (1995) 'Sustainability and community', *Australian Planner*, 32 (3): 109–15.

Jones, C. (1994) 'The future ain't what it used to be', *WTO News* (Madrid), 5: 12–13.

Long, V. (1993) 'Communities and sustainable tourism development', presented to 10th World Tourism Organisation Assembly, Bali.

Murphy, P. (1994) 'Tourism and sustainable development', in W. Theobald (ed.) *Global Tourism: The Next Decade*, Oxford: Butterworth Heinemann, 274–90.

Patel, S. (1993) 'Trends and challenges of tourism beyond the year 2000', *WTO Round Table on Beyond the Year 2000: Tourism Trends and Challenges*, Madrid: WTO, 10–13.

Pigram, J. (1991) *Outdoor Recreation and Resource Management*, London: Croom Helm (reprinted).

——(1993) 'Resource constraints on tourism: water resources and sustainability', 3rd Biennial Meeting, International Academy for the Study of Tourism, Seoul, June.

——(1995) 'Best practice environmental management and the tourism industry', 4th Biennial Meeting, International Academy for the Study of Tourism, Cairo, June.

Plimmer, N. (1993) 'Concluding Remarks', *WTO Round Table on Beyond the Year 2000: Tourism Trends and Challenges*, Madrid: WTO, 1–5.

Plog, S. (1972) *Why destinations rise and fall in popularity*, Los Angeles: Southern California Chapter, Travel Research Association.

Ritchie, B. (1993) 'Tourism research, policy and managerial priorities for the 1990s and beyond', in D. Pearce and R. Butler (eds) *Tourism Research: Critiques and Challenges*, London: Routledge, 201–16.

Strong, J. (1995) 'The Australian way', *Qantas Inflight Magazine*, July, 5.

Theobald, W. (ed.) (1994) *Global Tourism: The Next Decade*, Oxford: Butterworth-Heinemann.

Tourism Canada (1990) *Tourism Stream Conference: Action Strategy for Sustainable Tourism Development*, Ottawa: Tourism Canada.

Wahab, S. (1995) 'Tourism development in Egypt: competitive strategies and implications', 4th Biennial Meeting, International Academy for the Study of Tourism, Cairo, June.

World Commission on Environment and Development (1987) *Our Common Future*, Oxford: Oxford University Press.

World Tourism Organisation (1993) *Information*, newsheets, Madrid: WTO.

——(1995) 'Steady recovery in world tourism in 1994', *WTO News*, (Madrid) 1: 1.

3

SUSTAINABLE TOURISM – UNSUSTAINABLE DEVELOPMENT

Geoffrey Wall

The meaning of development and a number of competing development theories are reviewed. It is indicated that sustainable development has emerged as a concept into a field fraught with contention and debate. A distinction is made between sustainable development and sustainable tourism, and it is suggested that the latter in particular is an inadequate concept primarily because it is single-sector in orientation. Nevertheless, to the extent that the concepts engender a long-term perspective, foster notions of equity, encourage the search for and evaluation of types of tourism, promote an appreciation of the importance of intersectoral linkages, and facilitate dialogue between individuals and groups whose perspectives might at first sight appear to be at odds, they are useful catalysts in the search for more benign types of tourism which can contribute to long-term development, broadly conceived.

INTRODUCTION

Although claims are made that tourism is one of the largest, if not the largest, world industry and that international financial exchanges involved in tourism are comparable to those involved in oil and armaments, the most referenced document on sustainable development, *Our Common Future* (World Commission on Environment and Development 1987) does not even mention tourism. This is a major oversight and it reflects a lack of appreciation of the nature and significance of tourism.

Sustainable development is a political slogan rather than an analytical tool. It may be viewed as a philosophy, a plan or strategy, or a product. This is not to denigrate its utility in fostering dialogue among groups and individuals whose interests may at first sight appear to be incompatible. Thus it may be a catalyst for discussion, compromise and the identification of appropriate trade-offs between competing interests.

Sustainability has become a catchphrase which, partly because of its imprecision, has attracted widespread interest and support and also, again

33

partly because of its imprecision, criticism from its detractors. It has attracted the interest of the tourism industry and tourism researchers who have adopted the concept in slightly modified form and advocated 'sustainable tourism'. Such terminology implies a lack of appreciation of the full implications of the concept for it advocates a single rather than a multisectoral approach, the latter being a fundamental attribute of sustainable development. This is a particularly important problem in the case of tourism, which is a diffuse activity with far-reaching implications for many other sectors and activities.

While one can appreciate the interest of tourism industry advocates in ensuring the longevity of tourism, the goal should not be to perpetuate tourism at all costs. Rather, the question which should be asked is 'Whether and in what form might tourism contribute to sustainable development?'. This focuses attention not on tourism as an undifferentiated phenomenon but on types of tourism which are appropriate to particular situations, and the means for assessing and ensuring that they are sustainable.

THE NATURE OF DEVELOPMENT

Development is a slippery term. It is one which means different things to different people, and these meanings have changed over time. The term is value-laden, incorporating a mix of material and moral ideas encompassing both present and future states; what currently exists and how it came to be, as well as what might be brought into being in the future. In its early formulations it focused primarily upon economic matters. However, definitions have tended to be broadened over time and development has gradually come to be viewed as a social as well as an economic process which involves the progressive improvement of conditions and the fulfilment of potential. Now, in addition to economic issues, it encompasses social, environmental and ethical considerations and its measurement may incorporate indicators of poverty, unemployment, inequality and self-reliance (Binns 1995: 304). Notions of development have encompassed economic growth, structural change, autonomous industrialization and self-reliance, and have been pitched at a variety of scales from the individual to the regional, national and even international. However, growth and development are not to be regarded as the same thing, since economic growth as measured by gross national product per capita can occur simultaneously with increases in poverty, unemployment and inequality (Binns 1995: 305–6).

In its most basic form, development is concerned with human betterment through improvement in lifestyles and life opportunities. However, it is contentious as to how this is to be done and so the term has come to have strong political underpinnings. As is the case with sustainable development, the notion has also come to be used in rather different ways: as a philosophy, as a process, as a plan and as a product. As a philosophy, development refers to broad perspectives concerning appropriate future states and means of achieving them. As a process it emphasizes the methods which might be employed to expand or bring out the potentials or capabilities of a phenomenon. A development plan sets out specific steps through which desirable future states are to be achieved and development as a product indicates the level of achievement of an individual or society, as in developed, developing and underdeveloped countries. Clearly, these perspectives are not distinct but overlap to a considerable degree. Failure to acknowledge such differences in perspective can result in miscommunication.

The concept of development often has strong ideological underpinnings with conservative, liberal and radical traditions which have led to different perspectives on the causes of development challenges and their likely solutions. It also may have a built-in western bias as western societies are often seen as being developed in contrast to other countries which are seen as lacking in development.

As perspectives on development have changed through time, means of measuring development have evolved and broadened. Initial measures used were primarily economic, such as gross national product, or close derivatives such as energy consumption. But more recent trends have seen attempts to develop more comprehensive indices, including a variety of socioeconomic and environmental variables, even encompassing the extent of political and civil liberties and the quality of life.

Thus not only has there been an evolving debate on the nature of development, there have been changing perspectives on how to measure it. The evolution of research on tourism has paralleled the evolution of development studies as a whole, with an early emphasis on economic aspects now increasingly being complemented with a more balanced perspective incorporating environmental and sociocultural matters.

EVOLUTION OF DEVELOPMENT THEORIES

At the risk of oversimplification, four broad approaches to development can be identified which constitute a very rough evolution in the sequence of ideas concerning development. They are: modernization, dependency, neoclassical counter-revolution and alternative development. Each of these perspectives will be considered in turn. In reality there is considerable overlap both in the content of ideas and their timing so that there is a real risk of over-generalization. However, the ideas are presented simply in order to indicate the precursors to sustainable development. They have been informed particularly by the writing of Ingham (1993) and Goldsworthy (1988) and have benefited greatly from an unpublished synthesis by Telfer (1995).

Modernization

Modernization theorists have tended to view societies as passing through a series of development stages similar to those experienced by many western countries. Development has often been equated with growth emanating from relatively developed areas, and concepts such as stages of economic growth, growth poles, spread and backwash effects, and circular and cumulative causation have been guiding ideas with great stress being placed on the roles of innovation and entrepreneurship. State involvement to promote development has been encouraged and a belief in the trickledown effect, whereby benefits diffuse to disadvantaged people and regions, has been espoused. The Mexican strategy of building a number of very large resort complexes in such places as Cancun in the hope that the benefits will also accrue to a broader area, is an example of the application of this perspective. It can also be seen more generally in the proliferation of western-style hotels in relatively remote areas in the hope that they will be catalysts of economic growth.

The major criticisms of modernization theory are that it involves high levels of abstraction with limited discussion of the role of local involvement, that it suggests a unidirectional path which all must follow in order to develop, and that it smacks of western ethnocentrism as revealed in the First- and Third-World labels ascribed to parts of the globe. Modernization theory has little to say about the importance of traditional values, and perhaps implies that the maintenance of tradition and modernization may not be compatible goals.

Dependency

Dependency theorists see lack of development as being attributable to external forces more than internal causations, with power at the centre exploiting a disadvantaged periphery as described in centre–periphery models. For example, lack of development in developing nations may be seen as occurring as a result of exploitation by developed countries, often in the form of colonialism. These relationships may be viewed as being perpetuated by international tourism which may be viewed as a form of neocolonialism (Britton 1982, 1989; Nash 1989). Dependency theorists see a dualism between the rich and poor, the powerful and the powerless, both between and within countries, and see development as being best promoted by the favouring of domestic markets, import substitution, protectionism and social reforms.

Again, this is a fairly abstract and also rather pessimistic viewpoint. As already indicated, it stresses external relationships over internal problems and it tends to be somewhat vague on policy implications, tending towards protectionism and isolationism in contrast to modernization which would foster increased external economic links. Dependency theory is more a

critique of prevalent approaches to development than a method of development. In tourism, devotees of dependency theory have tended to spurn the
involvement of international corporations and capital and have advocated the
construction of small-scale, locally owned facilities, thereby overlapping
with the alternative development paradigm and alternative tourism which is
discussed below.

Neoclassical counter-revolution

Following the oil crises and the international debt crises of the 1970s and
80s, a group of ideas evolved which may be termed the neoclassical counter-
revolution. Proponents stressed the role of privatization and the free
competitive market. Development prospects were to be enhanced by
welcoming foreign investors with minimum state involvement. The World
Bank was a major proponent of this perspective, advocating structural
adjustments and reliance upon market forces and competitive exports. In
tourism, the World Bank, the United Nations Development Programme,
the Asian Development Bank and other similar international agencies have
helped to fund major tourism planning exercises in many parts of the
developing world.

The major criticism of this approach is that the financial strategies are
unlikely to help the disadvantaged who are most in need. Also, the scale of
some of the projects which were supported and the lack of detailed consideration of local conditions on the part of some of their advocates were a further
cause of concern.

Alternative development

Advocates of alternative development place emphasis on the satisfaction of
basic needs: food, housing, water, health and education. A grassroots
perspective is proposed with an emphasis on local involvement which, it is
argued, will permit people to control their own destinies. Sustainable development, with frequent associated demands for public participation and
community-level planning as advocated by Murphy (1985) for tourism, can
be viewed as an example of the alternative development paradigm.
Ecotourism, nature tourism, appropriate tourism, ethical tourism and
responsible tourism are but a few of the tourism descriptors which have
emerged in recent years, which have overlapping and imprecise meanings,
but which can all be grouped under the unenlightening heading of alternative tourism.

Critics of alternative development argue that the emphasis placed on
basic needs tends to detract from long-term growth, and that the stress on
local involvement underestimates the magnitude of political changes that
will be required to make this possible, with the paradox that political

changes could lead to too much state control. As indicated above in the form of sustainable development, the approach has been criticized as being too vague to be of much value.

Commentary

The ideas which have been presented above have strong political and ideological underpinnings (Goldsworthy 1988). Thus, for example, those with conservative dispositions have tended to advocate either open-market competition with minimal state involvement or, conversely, strong links between the state and private capitalists to promote top-down development. In contrast, liberals can be divided into non-structural reformists who have urged direct assaults on poverty and satisfaction of basic needs, and structural reformists who have sought broad-based changes, including reforms in the ownership and distribution of land in some circumstances, in order to promote a wider distribution of wealth and power. Marxism and Leninism are more radical approaches, the former seeing the class struggle as the route to development, whereas the latter places power in the hands of a political elite to organize production in the name of the people.

The above thumbnail sketches of development paradigms have been presented to indicate the contested and political nature of development and to show that sustainable development emerged as a concept into a field which was already fraught with contention and debate. However, before turning specifically to a consideration of sustainable development, two other sets of ideas will be presented briefly. The first is gender and development, which has evolved approximately contemporaneously with sustainable development, and the second concerns the evolution of tourism planning.

GENDER PERSPECTIVES

It is appropriate to consider gender perspectives at this point for a number of reasons. First, they provide a critique from a particular angle which is relevant to all of the development paradigms discussed above. They also reflect a stance which is slowly becoming a part of the tourism development literature. The viewpoints are also potentially compatible with sustainable development, for one of the tenets of sustainable development is equity and one of the objectives of the feminist literature is to draw attention to, and thereby help to redress, inequities associated with gender. In addition, it is a theme which reflects the fragmentation of the alternative development and alternative tourism literatures, but at the same time potentially cuts across them and could inform and be incorporated as a part of them.

Gender and development can be considered as a particular variant of alternative development. It is useful to make a distinction between sex and gender. Sex refers to biological attributes whereas gender refers to relationships between people of different sexes and has been defined by Henderson (1994) as 'a set of socially constructed relationships which are produced and reproduced through people's actions'. It follows that, in all but a very limited number of unusual cases, sex is fixed whereas the concept of gender is more dynamic, varying with changing interpersonal relationships. Nevertheless, there is an overlap between the two concepts if only because one's sex results in a lifetime of relationships and expectations based on and experienced as gender. Furthermore, gender is a cultural phenomenon, modification of gender relationships requiring changes in culture.

Major issues of concern in discussions of gender are access to power, control and equality. However, these do not depend solely upon gender. They are also influenced by such attributes as age, race, class, status and education. Thus, in discussions of gender, one should be aware of the dangers of single-variable analyses. There is no universal woman or universal women's experience but a variety of experiences which are influenced by many factors in addition to gender.

As a concept which stresses relationships, it should not be assumed that gendered approaches are only applicable to the study of women and their circumstances. In fact, gendered approaches can also be applied to men and a full gender analysis would involve investigation of both men and women and the relationships between them.

Gender scholarship

It is useful to provide a brief sketch of the history of gender scholarship. The ideas which are presented are drawn heavily from the writing of Swain (1995a). Gender scholarship is rooted in western feminism and its evolution over approximately the past thirty years. Five phases have been recognized:

1 in which women are essentially invisible and their experiences ignored (womanless);
2 a compensatory phase in which the experiences of prominent and unusual women were recognized but their activities were judged predominantly from a masculine perspective ('add women and stir');
3 an emphasis on dichotomous differences (sex differences). This is a useful starting point but it falls short if differences are left unexplained by reference to underlying circumstances;
4 feminist approaches which stress the experiences of women, often to the neglect or the exclusion of the experiences of men (women-centred);
5 true gender scholarship in which there is an interactional view of human expectations, behaviour and power relationships.

The above sequence broadly describes a path through which 'Women in Development' (WID) has given way to 'Gender and Development' (GAD), the latter having more explicit concern with the empowerment of women.

Tourism as a gendered phenomenon

Tourism and the processes associated with it emanate from gendered societies and are therefore likely to exhibit gendered relationships. Women are involved differently than men in both the production and consumption of tourism and so they are likely to be impacted differently. Relationships between gender and tourism are likely to be reciprocal, with gender influencing the manifestations of tourism and tourism modifying existing gender relationships. In addition, it is likely that gender relationships will differ between host and guest cultures and, furthermore, the behaviours of visitors and their expectations concerning gender relationships may differ when they are at home and on vacation. Thus the situation is extremely complex and, in the face of such complexity, both hosts and guests may fall back on stereotypes. The extent to which women and men are prepared to take up various positions, and the cultural acceptability of the employment of women and men in different aspects of tourism, will greatly influence the associated opportunities and impacts and their variation by gender, and thus the extent to which women and men participate in 'development'.

Investigators of tourism have not been in the forefront of researchers adopting a gender perspective, and in fact they have lagged behind their sisters in the broader field of leisure. Nevertheless, interest in gender and tourism is growing rapidly, as witnessed in the recent production of a number of documents on the topic, three of which will be referred to below. Norris and Wall (1994) prepared a literature review which examined the topic under the following six headings: the tourists (emphasizing the different experiences of women and men as tourists); tourism employment (stressing differences in employment opportunities with gender); types of tourism (on the basis that different types of tourism would impact upon women and men in different ways); images of tourism (particularly advertising); prostitution and tourism; and tourism and the family. Kinnaird and Hall (1994) recently published the first book on the topic. It consists predominantly of case studies but also includes a thoughtful introduction and a brief conclusion. Similarly, Swain (1995b) has edited a special issue of *Annals of Tourism Research* on 'Gender in Tourism' which, similarly, consists of an overview paper followed by a series of case studies. Together, these three sources provide a convenient introduction to a range of relevant topics and reflect the current status of research on gender and tourism. They reveal that much work is in the case-study mode and tends to be women-centred rather than 'true gender scholarship' as described above. However, the first tentative steps towards synthesis and conceptualization are being made.

It is pertinent to make a brief comment on the status of western and non-western scholarship. It appears that the feminist movement has received its primary impetus from western sources but that scholarship on gender and tourism, whether undertaken by western or non-western authors, has concentrated upon non-western settings. Thus, for example, the author is not aware of any studies of tourism undertaken in Canada which have adopted a gender focus. But there are a number of studies undertaken by Canadians in Indonesia which have a gender emphasis or component (Norris 1994; Cukier and Wall 1995), or have been undertaken from a Canadian base (Wilkinson and Pratiwi 1995).

Commentary

Gender and development offer a perspective which both competes with and is potentially compatible with sustainable development. To conclude this section, it is perhaps worth pointing out that sustainable development requires a sustainable labour force in appropriate numbers and with appropriate skills. Gender relationships underpin and have implications for virtually all aspects of tourism, including human resources development, product development, marketing, site and infrastructure planning, and impact assessment. Also, sustainable development requires that individuals, both women and men, be provided with opportunities to achieve their potentials and equity in access to those opportunities.

TOURISM PLANNING AND DEVELOPMENT PERSPECTIVES

Although a strong reason for encouraging investment in tourism, particularly in developing countries, is the expectation that it will contribute to development, there have been surprisingly few links established between the tourism and the development literature. Much of the economic literature as applied to tourism has been technical in nature, for example being concerned with the applications of cost-benefit analyses, economic multipliers and input-output analyses, and there has not been a strong proclivity to engage with the full range of debate on development perspectives as reviewed in the early part of this paper. Similarly, Ioannides (1995) has bemoaned the fact that there is a considerable gulf between the work of economic geographers and those geographers interested in tourism, although perhaps this is now beginning to change as exemplified in the recent writing of Brohman (1996) which touches on a variety of development issues.

Getz (1986) has provided a valuable review of the evolution of tourism planning traditions, and although his perspective is derived predominantly from western experiences, it reveals much about the ways in which tourism has been viewed as an agent of development. Getz recognizes four stages of tourism planning, followed by a fifth which is perhaps as much a goal as a stage which has been reached and superseded. Each stage is described according to basic underlying attitudes towards tourism, the problems which are addressed and the related methods and models employed by practitioners. Boosterism, which espouses an uncritical positive view of tourism and stresses exploitation of resources to attract more visitors, is replaced by a view of tourism as an industry comparable to other industries. The latter approach is predominantly economic in orientation and advocates the use of tourism to create jobs, earn foreign exchange, overcome regional disparities and modernize a society through the application of such development concepts as growth poles, and analytical techniques such as multiplier analyses and market segmentation. The physical/spatial approach sees tourism as a resource user and a threat to the environment and employs concepts such as carrying capacity, visitor management and impact assessment. It, in turn, is followed by community-based tourism which reflects a recognition that tourism is not inherently good or bad and that local control is desirable to direct the forms which tourism takes in the interests of community development. The fifth stage is called an integrative approach, in which a systems perspective is advocated and it is acknowledged that planning for tourism should be integrated with other planning processes.

The evolution of tourism planning, as described by Getz, exhibits many similarities to the sequence of development paradigms described above, in that a predominantly economic perspective is gradually replaced by one which incorporates a wider array of concerns and variables.

A CONTESTED CONTEXT

The preceding discussion has indicated that there has been considerable debate over the nature of development, how development should be measured and how it should be encouraged. This debate has strong ideological underpinnings. There is no widespread agreement on appropriate processes to promote development, or on the extent to which people should be the objects of development stimulated by others or the subjects of development relying primarily on their own initiatives and resources. In summary, there are varying perspectives on both objectives and processes; on what should be striven for and how it should be done. Thus sustainable development has arisen in large part because of dissatisfaction with existing perspectives; but it has come into being at a time of great uncertainty and has inherited many of the challenges of the development literature as a whole. The tourism literature and the practical application of tourism as a development tool have evolved with only a tenuous relationship to the development literature, although in spite of the lack of strong links there have been many parallels between them.

SUSTAINABLE DEVELOPMENT

Sustainable development can be viewed as a component of the alternative development paradigm discussed above. As in development as a whole, it may be regarded as a philosophy, a plan or strategy, or a product. It has been defined by the Brundtland Commission as 'development that meets the needs of the present without compromising the ability of future generations to meet their own needs' (World Commission on Environment and Development 1987: 43).

According to Wood (1993: 7), sustainable development has received widespread support because 'it appeared that sustainable development was an idea whose time had come, reflecting a convergence of scientific knowledge, economics, socio-political activity and environmental realities that would guide human development into the twenty-first century'. He suggests that it is a concept which acknowledges the needs of the world's poor and the limitations which are imposed on development by current levels of technical ability, social organization and environmental variability. It has received strong bureaucratic support at all levels, from local grassroots organizations to international agencies, partly because it reinforces a world view of economic growth as the engine of both development and environmental protection. Again, according to Wood (1993: 7) it lends legitimacy to the free-market economy, belief in trickledown economics and the benefits of technological progress. It offers a wide range of opportunities for action at all levels, and new institutions, policies and programmes which may be initiated under the guise of sustainable development provide significant opportunities to expand power bases, acquire additional resources and enhance prestige on the part of bureaucrats and administrators.

At the same time, sustainable development has been criticized on ideological grounds as promoting maintenance of the western capitalist system, as being too ambiguous and as a concept that tries to do something for everyone by tinkering at the margin of the economic system which originally created the problems that the concept is supposed to help address (Wood 1993: 8). At the extreme, one might regard sustainable development as an oxymoron: sustainable implying a state which can be maintained, is ongoing, perhaps even unchanging; whereas development implies change. Maintaining change, ongoing change or unchanging change are not clear messages! On the other hand, the ambiguity of the term potentially permits flexibility and fine-tuning to meet the needs of different places and cultures, encourages greater consideration of the environment, and more effectively integrates environmental and economic matters in decision making by encouraging dialogue between individuals with different perspectives. The concept is also in line with the community and participatory emphasis of much tourism literature. But unfortunately, the emphasis of many Third-World governments is not predominantly on the poor but on large-scale projects, both in tourism and other economic sectors, and generally provides little opportunity for local input. Thus the imprecision of the term has also been conducive to the formation of a gap between rhetoric and reality, and between policy and implementation.

Imprecision has also helped to disguise the fundamental re-orientation which is required in western thinking if sustainable development is to be sought and achieved. The scientific and technical approach which has resulted in great advances in knowledge has been achieved through the adoption of a narrow focus and reductionism. In contrast, sustainable development advocates holism and an appreciation of the interconnectedness of phenomena. Furthermore, humans have often been viewed as being separate from nature, which is there for humans to exploit, manage and control. Sustainable development implies that ultimately, humans and environments are indivisible. Conventional economics encourages people to act against their own best interests by placing greater value on present rather than future consumption. Growth in production and consumption is often equated with progress, and economic growth is still regarded as the best if not the only way to meet society's needs. Thus a belief in the value of modernity, the power of science and technology, and the law of supply and demand has been promulgated by western society and exported to other parts of the world. Such ideas do not fit snugly with notions of sustainable development.

SUSTAINABLE TOURISM AND SUSTAINABLE DEVELOPMENT

Widespread interest in sustainable development has captured the attention of tourism industry representatives and researchers and has resulted in the promulgation of sustainable tourism. Butler (1993: 29), who is one of the most articulate critics of sustainable tourism has defined it as follows: 'tourism which is in a form which can maintain its viability in an area for an indefinite period or time'. He contrasts this with a definition of sustainable development in the context of tourism as

> tourism which is developed and maintained in an area (community, environment) in such a manner and at such a scale that it remains viable over an indefinite period and does not degrade or alter the environment (human and physical) in which it exists to such a degree that it prohibits the successful development and well-being [sic] of other activities and processes.
>
> (Butler 1993: 29)

The distinction is critical and one which is not grasped by many advocates of sustainable tourism. It is essentially the distinction between a single-sector and a multiple-sector approach to development.

The first definition places the emphasis on the perpetuation of tourism to the neglect of other potential uses of scarce resources. However, tourism competes with other activities for the use of limited resources of land, water, labour and capital. The appropriation of resources may be in the narrow interests of the tourism industry, but may not be in the best interests of the broader community of interests. of which tourism is only a part. In fact diversity, whether it be in economy or biology, is likely to promote rather than detract from sustainability more broadly conceived.

The second definition acknowledges that tourism is unlikely to be the sole user of resources and that a balance must be found between tourism and other existing and potential activities in the interests of sustainable development. In other words, trade-offs between sectors may be necessary in the interests of the greater good.

But what is this greater good and what is to be sustained and who is to decide this? These are intractable questions. Should one be trying to sustain individuals, communities, regions or nations; experiences for tourists, incomes for businesses or lifestyles for residents; individual enterprises, economic sectors or whole economies and production systems; economic activities, cultural expressions or environmental conditions? Should all existing tourism developments be sustained or is it preferable that some be allowed to decline gracefully to be replaced by other activities? And should these new activities be touristic so that one could speak of sustainable tourism even though the form of tourism has changed and the new form might not contribute to the broader goal of sustainable development? Must all tourism developments be sustainable, or can one envisage situations in which tourism is advocated as a temporary, intermediate means to achieve other, long-term, goals? The examples of declining industrial centres, which have turned to tourism to upgrade their images and generate capital so that they can attract more investment of a variety of types, appear to suggest that the latter may be the case (Shaw and Williams 1994: 212–18).

If the goal for sustainable development is to meet human needs, then elaboration of these needs is required. Which specific needs are of concern? At what point are needs satisfactorily met? In the case of conflicting but valid needs, which needs should take priority and by what criteria is this to be determined? Is tourism to be regarded as a need of the present generation or is it an instrument which is to be employed to achieve sustainable development for other people at a future time?

The long-term perspective which is encouraged by the sustainable development doctrine is both a strength and a weakness. The strength is that it would support the precautionary and tie-in principles. The former advocates keeping opportunities open and not foreclosing options in the interests of short-term gains. The latter encourages taking actions, such as water conservation and the requirement that structures be set back from coasts, which make sense now but which will also increase resilience to future environmental perturbations. On the other hand, in the absence of a crystal ball, it is virtually impossible to determine which experiences will be desired by future tourists or residents, or to assess the influences of technological change for sustainability. How might one recognize a sustainable option in advance of the achievement of the associated sustainable state? And is what is sustainable now necessarily sustainable tomorrow given growing populations, emerging technologies and changing tastes?

The widespread adoption of the term sustainable tourism has been a mixed blessing. It has been adopted by the tourism industry to promote a clean and green image which is occasionally deserved, but more often is little more than a marketing gimmick. However, more positively, there are examples where transportation companies, hotels and restaurants have taken steps to recycle and reduce the consumption of energy, thereby reducing costs, increasing profits, and taking small steps in the direction of sustainability (Hawkes and Williams 1993). On the other hand, the interchangeable use of the terms ecotourism and sustainable tourism on the part of some spokespersons displays an inadequate understanding of both terms for, clearly, not all forms of ecotourism are sustainable and not all sustainable tourism need be to natural areas (Wall 1994). In fact, one might wonder if the average ecotourist is more demanding environmentally than the mass tourist who may not need to visit endangered species in remote locations, and whose needs and wastes can be more readily planned for and managed in large numbers incorporating economies of scale. Such questions cannot be answered with certainty until the research is done, but unfortunately the total costing of specific recreation activities or tourism types is an area which has yet to receive the attention it deserves.

In spite of the growing number of references to sustainable tourism, and even a journal which goes under that name (*Journal of Sustainable Tourism*) it is a moot question whether and in what form tourism can contribute to sustainable development (Wall 1993). The answers to such questions will likely be found in a greater appreciation of the consequences of different types of tourism. For this to occur, understanding of the implications of tourism must be enhanced (for example, of the nature of impacts and the role of scale of development), procedures must be devised to deal with complex decision making contexts (such as intersectoral linkages, top-down and bottom-up relationships, and coastal zone management), and improved procedures for implementation of plans and regulations must be devised (such as impact assessments and monitoring). Few would now argue that tourism is a pollution-free industry which is environmentally benign, although many would accept that it is in the long-term interests of the tourist industry to assure the longevity of the resources on which it depends. But this is easier said than done, for tourism exhibits many of the attributes of common property resources, with most money being spent on transportation, accommodation and food and beverages, and relatively little being directed to the maintenance of the natural and cultural heritage resources on which tourism ultimately depends. While most would agree that if tourism is to contribute to sustainable development it must be economically viable, environmentally sensitive and culturally appropriate, the forms which this might take are likely to vary with location. This in turn means that it will be difficult to come up with useful principles for tourism development which are true for all places and all times. Furthermore, the latter topic of culture is not well addressed in the literature on sustainable development, which has tended to focus upon tensions and compromises between economic development and environmental quality. Again, there are numerous questions and no easy answers.

CONCLUSIONS

Sustainable development is a form of alternative development which in turn is one among a number of development paradigms. The complexity of development paradigms has increased over time as narrow economic orientations have given way to more holistic perspectives incorporating a much wider range of concerns. Sustainable tourism, although rooted in the sustainable development paradigm, is an inadequate concept. While it has drawn attention to the need to achieve a balance between commercial and environmental interests, and has even spawned several successful examples of energy efficiency and recycling among tourist operations, as a single-sector concept it fails to acknowledge the intersectoral competition for resources, the resolution of which is crucial to the achievement of sustainable development. However, as Inskeep has suggested:

> the sustainable development approach can be applied to any scale of tourism development from large resorts to limited size special interest tourism, and that sustainability depends on how well the planning is formulated relative to the specific characteristics of an area's environment, economy, and society and on the effectiveness of plan implementation and continuous management of tourism.
>
> (Inskeep 1991: xviii)

Nevertheless, in spite of Inskeep's pertinent observations, the concepts of sustainable development and sustainable tourism both raise more questions than they answer. However, the questions are fundamental ones which require research and resolution. To the extent that the concepts engender a long-term perspective, foster notions of equity, encourage the search for and evaluation of types of tourism, promote an appreciation of the importance of intersectoral linkages, and facilitate dialogue between individuals and groups whose perspectives might at first sight appear to be at odds, they are useful catalysts in the search for more benign types of tourism which are likely to contribute to long-term development, broadly conceived. However, although favoured by industry advocates, sustainable tourism is a less desirable term with a slightly different message than sustainable development. The quest for sustainable tourism may be sufficient to meet the narrow interests of the tourism industry but the search for sustainability more broadly conceived, in which the tourism industry may be a partner and in which tourism is viewed as a means rather than an end, is likely to address more fundamental development goals. Just as tourism is rarely the sole cause of the problems for which it is sometimes indicted, it is unlikely to be the complete solution to all development needs.

REFERENCES

Binns, T. (1995) 'Geography in development: development in geography', *Geography*, 80: 303–22.

Britton, S. G. (1982) 'The political economy of tourism in the third world', *Annals of Tourism Research*, 9: 331–58.

——(1989) 'Tourism. Dependency and development: a mode of analysis', in S. G. Britton and W. C. Clarke (eds) *Ambiguous Alternative: Tourism in Small Developing Countries*, Suva: University of the South Pacific.

Brohman, J. (1996) 'New directions in tourism for third world development', *Annals of Tourism Research*, 23: 48–70.

Butler, R. (1993) 'Tourism: an evolutionary perspective', in J. G. Nelson, R. Butler and G. Wall (eds) *Tourism and Sustainable Development: Monitoring, Planning, Managing*, Department of Geography Publication series no. 37, Waterloo, Canada: University of Waterloo, 27–43.

Cukier, J. and Wall, G. (1995) 'Tourism and employment in Bali: a gender analysis', *Tourism Economics*, 1: 389–401.

Getz, D. (1986) 'Models in tourism planning: towards integration of research and practice', *Tourism Management*, 17: 21–32.

Goldsworthy, D. (1988) 'Thinking politically about development', *Development and Change*, 19: 505–30.

Hawkes, S. and Williams, P. (eds) (1993) *The Greening of Tourism from Principles to Practice: A Casebook of Best Environmental Practice in Tourism*, Burnaby: Centre for Tourism Policy and Research, Simon Fraser University.

Henderson, K. (1994) 'Perspectives on analyzing gender, women and leisure', *Journal of Leisure Research*, 26: 119–37.

Ingham, B. (1993) 'The meaning of development: interactions between "new" and "old" ideas', *World Development*, 21: 1803–21.

Inskeep, E. (1991) *Tourism Planning: An Integrated and Sustainable Development Approach*, New York: Van Nostrand Reinhold.

Ioannides, D. (1995) 'Strengthening the ties between tourism and economic geography: a theoretical agenda', *The Professional Geographer*, 47: 49–60.

Kinnaird, V. and Hall, D. (eds) (1994) *Tourism: A Gender Analysis*, Chichester: Wiley.

Murphy, P. (1985) *Tourism: A Community Approach*, London: Methuen.

Nash, D. (1989) 'Tourism as a form of imperialism', in V. Smith (ed.) *Hosts and Guests: The Anthropology of Tourism*, Philadelphia: University of Pennsylvania Press, 37–52.

Norris, J. (1994) *Gender and Tourism in Rural Bali: Case Study of Kedewatan Village*, University Consortium on the Environment, Student Paper no. 22, Waterloo, Canada: University of Waterloo.

Norris, J. and Wall, G. (1994) 'Gender and tourism', *Progress in Tourism Recreation and Hospitality Management*, 6: 57–78.

Shaw, G. and Williams, A. (1994) *Critical Issues in Tourism: A Geographical Perspective*, Oxford: Blackwell.

Swain, M. (1995a) 'Gender in tourism', *Annals of Tourism Research*, 22: 247–66.

Swain, M. (ed.) (1995b) 'Gender in tourism', *Annals of Tourism Research*, 22: 2 (special issue).

Telfer, D. (1995) 'The integration of development theory and tourism research', mimeo, Waterloo, Canada: University of Waterloo.

Wall, G. (1993) 'International collaboration in the search for sustainable tourism in Bali, Indonesia', *Journal of Sustainable Tourism*, 1: 38–47.

——(1994) 'Ecotourism: old wine in new bottles?', *Trends*, 31, 2: 4–9, 47.

Wilkinson, P. and Pratiwi, W. (1995) 'Gender and tourism in an Indonesian village', *Annals of Tourism Research*, 22: 349–66.

Wood, D. (1993) 'Sustainable development in the third world: paradox or panacea?', *The Indian Geographical Journal*, 68: 6–20.

World Commission on Environment and Development (1987) *Our Common Future*, Oxford: Oxford University Press.

4

MASS TOURISM

Benefits and costs

Norbert Vanhove

Mass tourism is related to two main characteristics: (a) participation of large numbers of people in tourism; and (b) the holiday is standardized, rigidly packaged and inflexible. The number of international tourist arrivals is expected to continue to grow during the next decade, and with it the phenomenon of mass tourism. Benefits and costs of tourism can be measured at different levels: national, regional or local. In all cases a social cost-benefit analysis is the adequate approach. In such an approach paid and unpaid benefits and costs and side effects are taken into account. The key benefits of mass tourism are income and employment generation. For both benefits input-output analysis is the best method of assessment. The key cost items are the so-called incidental costs. These lead to quality-of-life costs and public or fiscal costs. To cope with the negative impacts, attention should be paid to (a) staggering of holidays in time, space and product; (b) tolerable numbers as a central issue in tourism planning; and (c) a better behaved kind of tourist. For sustainable tourism a region should put environment first. This means building responsible tourism, fostering a culture of conservation and developing an environmental focus.

MASS TOURISM

Definition of Mass Tourism

Mass tourism is a notion in common use. But what does it mean exactly? Is it a package tour? Is it a concentration of tourists in a resort or region? Is it tourism with a low profile? These are only some aspects of the phenomenon.

According to the Swiss author Fink (1970) the basic elements of mass tourism are:

- participation of large numbers of people;
- mainly collective organization of travelling;
- collective accommodation;
- conscious integration of the holidaymaker in a travelling group.

A more general and workable definition was found with Burkart and Medlik:

Mass tourism refers to the participation of large numbers of people in tourism, a general characteristic of developed countries in the twentieth century. In this sense the term is used in contrast to the limited participation of people in some specialist forms of tourist activity, such as yachting, or in contrast to the situation in developing countries or in countries with extreme inequalities of income and wealth or, indeed, to the limited extent of tourist activity everywhere until a few decades ago. Mass tourism is essentially a quantitative notion, based on the proportion of the population participating in tourism or on the volume of tourist activity.

(Burkart and Medlik 1974: 42)

Poon (1993: 32) emphasizes more the large-scale packaging of standardized leisure services, and for her mass tourism exists if the following conditions hold:

- The holiday is standardized, rigidly packaged and inflexible. No part of the holiday could be altered except by paying higher prices.
- The holiday is produced through the mass replication of identical units, with economies of scale as the driving force.
- The holiday is mass-marketed to an undifferentiated clientele.
- The holiday is consumed *en masse*, with a lack of consideration by tourists for local norms, culture, people or the environments of tourist-receiving destinations.

The notion of mass tourism should be distinguished from the notion of popular and social tourism. Once again to refer to Burkart and Medlik:

Popular tourism denotes tourist activities meeting with a wide acceptance by people, because of their attractiveness and availability. The acceptance may be due to meeting the needs or tastes of people or more particularly to being available at a low price. Popular tourism is, therefore, essentially a qualitative notion, although by its nature it may give rise to mass tourism.

(Burkart and Medlik 1974: 43)

As distinct from the former two examples, social tourism is concerned specifically with the participation in tourism of people of limited means and with the measures to encourage this participation and to make it possible.

From these definitions two main features can be derived:

- the participation of large numbers of people in tourism, whatever the tourist activity may be;
- the holiday is mainly standardized, rigidly packaged and inflexible.

51

NORBERT VANHOVE

Participation in practice

The phenomenon of mass tourism can be illustrated with the evolution of the net and gross participation rate of the residents of a number of European countries. Table 4.1 and Figure 4.1 show the evolution of the holiday participation rate (four nights and more) for Belgium, France, The Netherlands, The United Kingdom and West Germany (Boerjan and Lowyck 1995). All five countries are important generating markets. The data for all these

Table 4.1 Holiday participation in a number of European countries

Year	Belgium[a]		Great Britain[b]		Netherlands[c]		France[d]		West Germany[e]	
	A	B	A	B	A	B	A	B	A	B
1967	34	42								
1969							45	68	42	49
1970			59	75						
1975			60				53	84		
1976	47	63								
1980			62	89	68	97	56	101	58	70
1982	48	64					58	106		
1985			58		64	92	58	106	57	67
1986			60	89	65	95	58	109	57	67
1987			58	88	67	108	59	113	65	78
1988	56	83	61	97	70	111	60	115	65	84
1989			59	94	72	119	61	117	67	88
1990			59	95	70	117	59	114	68	84
1991	61	97	60	96	71	120	60	116	67	82
1992			59	96	70	117	60	116	71	90
1993			61	100	71	123	61	119	75	101
1994	63	100	60	102	73	126	62	na	79	108

A: Net participation
B: Gross participation
a: WES, 'Reisgedrag van de Belgen'; the method of inquiring is not fully comparable, throughout the period considered.
b: BTA/ETB, 'British National Travel Survey'.
c: NRIT/NBT, 'Continu vakantie-onderzoek'.
d: INSEE, 'Les vacances des Français'.
e: DIVO Reiseanalyse, 'Studienkreis für Tourismus, Reiseanalyse und FUR Urlaub und Reisen'.
Source: Boerjan and Lowyck 1995

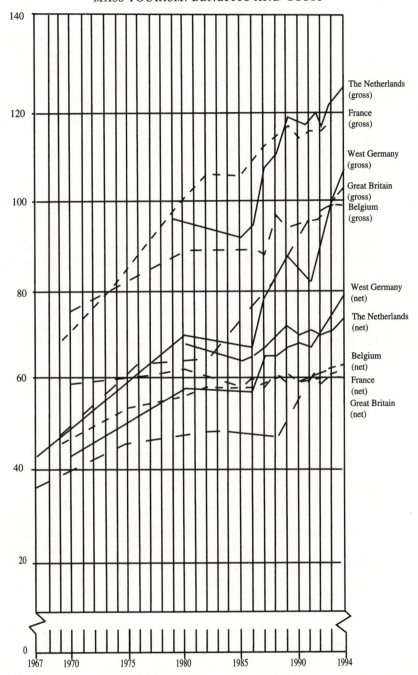

Figure 4.1 Evolution of net and gross participation rate
Source: Westvlaams Ekonomisch Studiebureau

countries are not fully comparable, for example, the definition of a holiday is not the same in all countries (e.g. holidays in second homes and holidays with family and friends). Furthermore, the population is differently reckoned: it can mean total population, or people of fourteen years and older.

Table 4.1 and Figure 4.1 illustrate three important facts. First, the net participation rate (number of holidaymakers per 100 inhabitants) in the retained generating markets is very high. For 1994 the rate varies between:

Great Britain: 60.0
France: 62.0
Belgium: 63.2
Netherlands: 73.2
West Germany: 78.6

Second, during the last three decades there has been a substantial increase in the number of holidaymakers. However, in two of the five countries, the net participation rate seems to have reached a ceiling. For the period 1970–94 the net participation rate in Great Britain is close to sixty. The same trend for France for the period 1985–94 is noted, and The Netherlands and West Germany may also have reached a maximum or the ceiling.

Third, the main growth of the holiday market can be noticed in the gross participation rate. For 1994 the number of holidays per 100 inhabitants varies between:

Belgium: 100.4
West Germany: 107.7
France (1993):   119.3
Netherlands: 125.9

It is difficult to predict the ceiling for the second and third holidays, although it appears to be still far from a maximum.

Night participation

It is, however, wrong to believe that the accommodation sector underwent a similar increase in demand for lodging capacity. In other words nights stayed did not increase to the same extent. On average the holidays became shorter. This can be illustrated with the night participation rate in a number of countries. The night participation rate stands for the number of nights per inhabitant (here limited to holidays).

Table 4.2 illustrates that the increase in the night participation rate is far lower than the increase in holiday participation during the period 1980–93. France has experienced a stagnation of the night participation rate. In the Netherlands the rate has still increased, from 13.0 in 1980 to 14.7 in 1994. This growth is, however, far below the evolution of the holiday participation rate. On the other hand Belgium has seen a substantial increase in night

Table 4.2 Night participation rates in a number of European countries

Year	Belgium	France	Netherlands
1969	na	13.7	na
1976	10.0	na	na
1980	na	16.6	13.0
1982	9.9	na	na
1985		16.8	11.8
1988	12.3		13.5
1990		16.5	13.8
1991	13.2		13.9
1993		16.4	14.2
1994	13.6	na	14.7

Source: Westvlaams Ekonomisch Studiebureau

participation. Belgium is, however, the country (in Table 4.1) with the highest increase in holiday participation.

The different evolution of gross holiday participation and the night participation rate has an unequal impact on the subsectors of tourism such as transport, travel agencies, tour operators and ingoing tour operators on the one hand, and accommodation on the other.

Short holiday participation

A new aspect of the holiday market is short holidays. Indeed, besides holidays (four nights and more), more attention should be paid to short holidays of between one and three nights. The growth of short holidays is at least surprising and more spectacular than the expansion of second holidays (see Table 4.3). Also, here a simple comparison between countries is not valid. In certain countries (e.g. Belgium) second residences and holidays with family and friends are not considered. In other countries (e.g. The Netherlands and West Germany) only holidays with family and friends are excluded.

Table 4.3 Evolution of net and gross participation rate of short holidays in a number of European countries

Year	Net participation			Gross participation		
	Belgium	Netherlands	West Germany	Belgium	Netherlands	West Germany
1980	na	21.0	25.7	na	49.4	56.5
1982	10.9	na	na	17.4	na	na
1985	13.0	19.6	32.6	16.7	44.1	65.7
1988	18.5	29.2	36.4	26.3	69.5	88.6
1990			37.0			79.5
1991	24.1	32.4	na	34.6	70.3	na
1993		34.7	40.9		71.1	98.2
1994	27.8	34.4	41.4	39.9	76.7	90.8

Source: Westvlaams Ekonomisch Studiebureau

From Table 4.3 it can be seen that the short holiday market is booming, and the rates for these three countries are typical for all Western European countries.

One conclusion is evident. Tourism is a mass phenomenon. For most Europeans a holiday each year is a 'must', even when the economic situation is temporarily less brilliant. Economic crisis may mean the destination is different or the stay shorter, but none the less the participation rate is more or less unaffected.

International tourist arrivals

So far the discussion has dealt with participation rates in generating countries. On the other side of the coin are the arrivals in receiving countries. Of course not all holidays are taken abroad; a high percentage of tourism demand is domestic. Figure 4.2 shows the evolution of international tourist arrivals during the period 1960–94, and in Table 4.4 one can find the average growth rate of arrivals per region and sub-period.

Three points seem to be evident:

1 International tourism has grown spectacularly over the last three decades. In fact the phenomenon started after the Second World War.
2 This growth is very unequally spread over the different regions. East Asia and the Pacific have become the new receiving areas of the world.
3 There is a trend to lower growth rates in the last decade, especially during the first half of the 1990s.

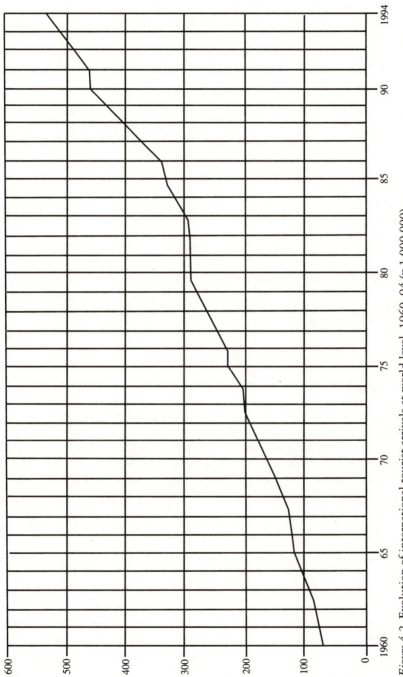

Figure 4.2 Evolution of international tourist arrivals at world level, 1960–94 (x 1,000,000)
Source: WTO 1995

Table 4.4 Annual growth rate of tourist arrivals per region and period, 1960–94 [a]

Region	1960–70	1970–80	1980–90	1990–94
Africa	12.3	11.7	7.4	4.9
America	9.7	3.7	4.3	3.2
East Asia and the Pacific	22.4	14.6	9.7	8.8
Europe	8.4	5.1	4.2	3.1
Middle East	11.4	12.3	2.3	2.1
South Asia	17.6	9.5	3.3	3.8
World	9.1	5.5	4.8	3.9

a: Annual growth rate must not be confused with average annual percentage increase as published by World Tourism Organisation.

Source: WTO, *Yearbook of Tourism Statistics*, 1995

In his publication *Trends in International Tourism*, Baum (1995) underlines a number of key factors for the growth of mass international tourism:

- economic growth;
- more leisure time;
- lifestyle and work-related changes;
- technological change (especially in aviation and communications); the automobile and the aeroplane become reliable modes of travel for a large middle class;
- changes in the demographic structure of the most affluent societies, especially improved health among the retired;
- reduced international tensions; less bureaucracy involved in travel to and from many countries; and
- a growing awareness of and interest in other cultures and ways of living, a sense of belonging to 'one world'.

What about the future?

Today the number of international tourist arrivals is close to twenty times what it was in 1950. Can this trend be expected to continue? The last long-term travel forecasts, the fifth set published by the Economist Intelligence Unit (EIU), leave no doubt. Travel is projected, in terms of trips made, from each of the world's thirty leading countries of origin to each of ten standardized major destination regions (e.g. the Caribbean, with twenty-nine destinations), of which the Europe/Mediterranean region is by far the most important (Edwards 1994).

In aggregate, the EIU's forecasts expect world tourism to grow significantly more slowly in 1989–95 than in the 1980s. This is largely, but not solely, due to the impact of the 1991–92 economic recession. During 1995–2000 there should be an appreciably faster growth, but with some slowing after 2000. During 1989–2005 as a whole the forecast overall growth rate of 4.2 per cent per year in terms of trips is expected to be very similar to that experienced during 1980–89. As Table 4.5 shows, throughout the forecast period long-haul travel is projected to grow substantially faster than short/medium-haul journeys.

In 1995 the World Tourism Organisation (WTO) published its own global tourism forecasts. They are in line with the projections of Edwards (1994). For the period 1990–2000 WTO predicts a worldwide growth rate of 3.8 per cent, and for the next decade, 2000–10, 3.6 per cent. This would lead to 661 million international arrivals in 2000 and 977 million in 2010.

When the origin-destination flows published in the above-mentioned EIU study are analysed carefully – and WTO publications confirm this – almost all regions of the world are involved in the phenomenon of mass tourism. The economic impact must be great and has two different aspects: benefits and costs.

BENEFITS

The misleading literature of the 1970s

From the above it is clear that mass tourism is not a phenomenon of the 1990s, but started in the 50s and 60s. Nevertheless it is surprising that at the first Lomé conference in 1975 (of less developed countries), tourism was not considered as a sector supporting economic development of ACP countries (Africa, the Caribbean and the Pacific). Lomé II paid a little attention to it. It was only in 1985 on the occasion of Lomé III that tourism received the interest it deserved.

Table 4.5 Forecast annual growth rates in world travel, 1989–2005 (in %)

Trips	1989–95	1995–2000	2000–5
Short/medium haul	3.3	4.8	3.6
Long haul	4.7	6.8	6.1
All trips	3.5	5.1	4.0

Source: Edwards A., 'International tourism forecasts to 2005', The Economist Intelligence Unit, 1994

Why was it so long before tourism was recognized as a valuable component of economic development? In the 1970s many publications – reports, books, and articles – were written by authors who had never seen a developing country and/or only emphasized the negative effects of tourism due to insufficient economic knowledge. Import leakages, income transfers, foreign ownership, tourism as a factor of inflation, destruction of culture, mono-industry and social impacts were keywords of this literature. It cannot be denied that all these negative factors more or less exist in many destinations. However, from an economic point of view it is not realistic to deny the positive factors. One cannot imagine mass tourism without economic return.

Many benefits are mentioned in the literature. Unfortunately this is a phenomenon as chaotic as the Tower of Babel. Variables frequently mentioned are expenditure, income generation, employment creation, foreign exchange earnings, tax receipts, social benefits, tourism multiplier, transaction multiplier and many more benefits or presumed benefits. Very often these variables are not put in their right context or relationship.

In this section the benefits of mass tourism for a nation, region or local community are considered. Further, the benefits of a tourist project within the framework of a cost-benefit analysis are examined.

Benefits for a nation, region or town

The two basic incentives for the tourism sector are expenditure and investment. Both are essential elements of final demand in an input–output table. Receipts in foreign exchange are a component of expenditures and as such the export element of final demand. Foreign exchange earnings in tourism can be of great importance in certain countries. Spain is a good example. Some developing countries can improve their international liquidity position with earnings from international tourism.

The basic elements of the final demand, tourism expenditure – tourism export included – and tourism investment, are the bases of two key benefits: income and employment. In other words what does tourism create in terms of income and employment generation? In this way the relationship between four key variables in the tourism sector is linked: expenditure and investment on the one side and income and employment on the other side.

Between these four key variables there are special links. Investments depend on the expenditures in the present, the past and the future. In turn, investments can stimulate expenditures.

What is the relationship of the key variables with other above-mentioned variables? Starting with government receipts, the latter are a derivative. In principle, each economic activity can yield returns for the public sector such as direct taxes, value-added tax (VAT), company taxes, social security receipts, etc. These returns are a function of the key variables, as is the case for any other economic sector.

Tourism as a sector does not have an independent existence in society. Each tourism phenomenon provokes social costs and social benefits. This can be illustrated by an important investment such as a holiday village or a theme park, products of mass tourism. Such an investment is responsible for social costs which are not paid by the investor. The same projects lead to social benefits which do not accrue to the investing company. One group of social benefits are the well known tourism multiplier effects. On this subject, Archer (1991) did a lot of interesting research. The multiplier mechanism works as follows (see Figure 4.3).

Taking the expenditure in a hotel as a starting point, to whom does this expenditure accrue? One part creates value-added or factor remuneration in the hotel. It is direct income within the region concerned. A second part leads to local business transactions. A hotelier must restock inventories to provide for future sales (bread, meat, vegetables, fruit, etc.). A third part of the expenditure is used to pay profit taxes, local taxes, etc., to local, regional or national governments. A fourth part is spent on leakages such as imports of goods (e.g. whisky) and payment of profits to people and organizations outside the region or country.

Purchases of groceries provoke in turn the same above-mentioned effects – income creation, intermediate purchases, public transfers and import leakages – with the butcher, the baker, the farmer, and so forth. This process continues with a third and a fourth round. After each round the national or regional effects become smaller. The income created in the successive rounds is called the indirect income. The degree of magnitude of these indirect effects is governed by the extent to which business firms in the nation or region supply each other with goods.

However, a second derived impact can be noticed. The more that wages and profits (direct and indirect) due to the hotel expenditure rise, the more consumer expenditure increases and this provides a further impetus to economic activity. Additional business turnover occurs and this generates income. These are the so-called induced effects.

The indirect and induced effects can be quite considerable in the absence of important leakages, such as savings (assuming that sufficient resources are available).

One key observation should be made. There are few arguments to suggest that induced effects in tourism are different from the same effects in other sectors. Consumer behaviour of tourism earners cannot be so different from textile earners or other economic sectors.

This multiplier mechanism leads to different types of multipliers (e.g. orthodox and unorthodox).

Archer and Fletcher (1990) have compiled a number of orthodox income multipliers (direct + indirect + induced income creation over direct income creation) from reports and publications. The calculation of these multipliers

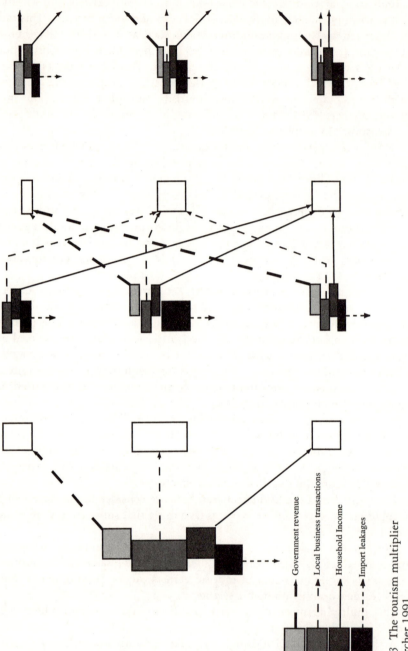

Figure 4.3 The tourism multiplier
Source: Archer 1991

Government revenue
Local business transactions
Household Income
Import leakages

is based on different methods and refers to different periods. Some of them are selected to illustrate how different they can be.

Turkey	1.96
United Kingdom	1.73
Ireland	1.72
Egypt	1.23
Bermuda	1.17
Missouri (USA)	0.88
Gwynedd (UK)	0.37
East Anglia (UK)	0.34
City of Winchester (UK)	0.19

It is obvious that the larger the area is, the higher the corresponding tourism multiplier value. The value of the multiplier is determined by the structure of the economy, the intersectoral relations, the import content, the nature of the tourism product, etc.

In the first place the interest is in the unorthodox income multiplier of the type direct + indirect income creation over tourist expenditure. With the knowledge of the expenditure and the tourism income multiplier (TIM), direct and indirect income creation can be estimated. The crucial point is the knowledge of TIM, and that is unknown. However, one can often make use of TIM values for countries and/or regions with similar products and general common circumstances.

What is now the relationship of all this to mass tourism? The linkage is evident. The WTO estimated for 1994 the receipts of international tourism at about US$336 billion (international transport excluded). Mass consumption leads to a high income and employment generation.

Income and employment generation

The importance of income and employment generation can be illustrated by case studies. The first one is Flanders (Vanhove 1993). Flanders, part of Belgium, has besides an industrial tradition a number of famous attractions (art cities and the coastal area). The second case is the Seychelles.

In 1991 Flanders' tourism expenditures were equal to 159 billion BF (Belgian Francs) (1US$ = ± 30BF) (domestic and international tourism). This was the starting point to estimate the income generation using several methods. The most correct one – if data are available – is input-output analysis. The implementation of this method implies four steps:

1 definition of final production;
2 breakdown of tourism expenditure to input-output sectors;
3 estimation of direct income creation:
$$Y_d = \hat{B}_k \cdot b$$

in which:

Y_d = direct income generation

$\hat{B}_k$ = diagonal matrix of income coefficients

b = column vector of tourist expenditure after elimination of imports and VAT

4 estimation of indirect income generation:

$Y_t = \hat{B}_k (1 - A^n)^{-1} \cdot b$

in which:

Y_t = direct and indirect income creation

$(1-A^n)^{-1}$ = inverse matrix

The application of this method resulted in a direct income creation of BF75 billion and BF26 billion indirect income creation.

Employment creation was calculated in a similar way:

$$E_t = \hat{A}_k (1 - A^n)^{-1} \cdot b$$

in which:

E_t = vector with direct and indirect employment

$\hat{A}_k$ = diagonal matrix of employment coefficients

This application resulted in direct employment of 54,900 and indirectly of 19,000 people. This employment effect is of the same magnitude as most important industrial sectors in Flanders.

With the same method it is relatively easy to calculate the direct and indirect government receipts, as shown in Table 4.6. In other words an expenditure of BF159 billion creates BF41.7 billion in government receipts.

A similar and interesting piece of research was carried out by Archer and Fletcher for the Seychelles (Archer and Fletcher 1996) also using the input-output model. However, in their application, induced effects were retained.

Table 4.6 Direct and indirect effects

	Direct + indirect effects
Direct taxes	7.1 billion BF
Profit taxes	5.6 billion BF
Indirect taxes	4.0 billion BF
Related to production VAT	14.2 billion BF
Social security	10.8 billion BF
Total	41.7 billion BF

This study proves how important tourism is in terms of income and employment creation. The reference year was 1991, and the case is relevant for the impact tourism can have.

The major data of economic impact for the Seychelles are:

- visitor arrivals: 97,668
- tourist nights: 938,000
- tourism expenditure: 527 million SEYRs (currency unit for the Seychelles) or US$99 million
- income creation: 466 million SEYRs (direct, indirect and induced)
- government revenue: 148 million SEYRs
- direct employment generated: 3,772
- total employment generated: 8,312 (direct, indirect and induced)

The contribution of tourism to GDP was 18.4 per cent. With the secondary effects resulting from the multiplier action, however, tourism contributes approximately 23.5 per cent to GDP.

About twenty-four tourists create one direct job, but with secondary jobs taken into account as well, only 10.8 tourists were needed to support one job. In terms of nights, 248 nights contribute to a direct job and with secondary jobs 113 nights are sufficient for one job. A similar figure for Bruges (948 nights for a direct job) shows the very labour-intensive character of tourism in Seychelles.

Of course one cannot generalize the results for Flanders and the Seychelles to any country or region. All depends on leakages, economic structure, tax systems, etc. Nevertheless, one can state that, on average, the impact of mass tourism on income and employment generation must be huge. This conclusion also holds for most developing countries. The income generation might be lower but very often the employment generation is much higher.

Special characteristics of employment

So far the quantitative employment effect of tourism has been stressed. Besides this aspect, one should pay attention to the qualitative aspects.

First, tourism is a growth sector and all predictions for the next decade – EIU and WTO – are very optimistic. Even in developed countries, tourism is a sector with promising job opportunities.

Second, tourism is a sector with a high degree of semi-skilled and so-called unskilled employees. This rather negative aspect can be seen as an opportunity for the large number of unskilled workers with a job, especially in developing countries.

Third, tourism is a sector with a high percentage of part-time jobs. Hudson and Townsend reveal that in Great Britain 38 per cent of the men and 56 per cent of the women working in the hospitality sector (hotels, restaurants and cafes) are part-time workers (Hudson and Townsend 1992).

According to Wood there is in advanced industrial societies a tendency to increased part-time employment and the casualization of work (Wood 1992).

A fourth characteristic of employment in tourism is the high share of female workers. This share amounts to 45 per cent among full-time workers and 73 per cent among part-time jobs.

Fifth, the sector has many small firms and self-employed.

Sixth, British sources indicate very often the increasing number of young workers in the tourism sector (Wood 1992; The Host Consultancy 1991).

A final characteristic is the seasonality of employment. The intensity of seasonality is different per country and tourism region. Where it exists, it makes tourism employment less attractive.

All in all, employment in tourism is growing, but most characteristics point to the fact that the sector has a low image.

Special benefits for less developed countries or regions with tourism resources

A tourist product is composed of several elements: attractions, facilities, transport, entertainment, image, etc. The basic element is the attractions. These can be of very different types. Many developing countries or developing regions in Europe are rich in natural or human-made attractions. A development based on these attractions offers the tourism sector some comparative advantage *vis-à-vis* other economic sectors.

The first comparative advantage is directly related to natural attractions (e.g. sun, beaches, mountains, etc.). These attractions are raw materials which can be made beneficial at reduced costs, and the danger of exhaustion is more or less non-existent. Here reference is made to an old but very good contribution of Mossé:

> Besides, the host country may have been endowed by nature with an abundance of readily marketable assets for whose enjoyment tourists are willing to pay: sandy beaches, picturesque sites (mountains and forests), a sunny climate, and the remnants of ancient civilizations. Out of the 20 dollars, the tourist may well have gladly spent 5–6 dollars to enjoy these 'free utilities', as Bastiat would have called them, on whose supply the host country did not have to spend a penny, either in local or in foreign currency.
>
> (Mossé 1973: 31)

The second comparative advantage concerns the import content. There are good grounds to believe that tourism on average has a lower import content than other basic economic sectors. A number of publications support this thesis (Theuns 1975; United Nations Centre for Tourism and Development (UNCTAD) 1971). The reason is evident. The tourists are buying services which the local population can to a large extent provide. Furthermore, it is not too difficult – at least in most regions – to develop in the long run the agricultural sector towards the needs of the hospitality industry. Mossé supports this point of view. 'As a source of foreign exchange, tourism is on a par with other export industries, but with one difference: they (export industries) require costly inputs' (Mossé 1973; Vanhove 1977).

The third advantage is the very high growth rate noted earlier. This growth, together with the good perspective and high income elasticity, makes tourism a preferential sector for economic development.

Fourth, tourism has a stabilization effect on exports. Export markets in raw materials are unstable, and therefore foreign earnings are uncertain. The price obtained for raw materials is governed by the world market price and is subject to terms-of-trade conditions. To avoid a deterioration of terms-of-trade, tourism development is often a solution. Mass tourism yields important amounts of foreign exchange, which allow the country to import manufactured goods. The counterpart is a limited quantity of resources. To quote Mossé:

> A balance of what is given and what is received should be struck not on the basis of the hours of work necessary, but in regard to the utility of the items exchanged. The utility of exports consisting of abundant wild fruit, stretches of sandy desert or trees growing by themselves in the forest, is insignificant, in contrast to the great utility of importing electrical and telephone equipment, transport facilities, etc.
>
> Unfortunately, there is no universally accepted measure of utility and use must be made of costs expressed in money or manhours.
>
> (Mossé 1973: 33)

But the tourism sector should not be presented as too optimistic a matter. Tourism is very sensitive to internal problems, political events, diseases, bad news, etc.

A fifth comparative advantage is related to the labour-intensive nature of the sector. This high labour-intensity is notable in the accommodation sector, the subcontracting sector, services, etc. This comparative advantage finds a lot of support in economic theory.

Other benefits

Development of tourism on a large scale, based on mass tourism, creates external economies. Improvements in transportation networks, water quality and sanitation facilities may have been prompted by the tourist industry but benefit other sectors of the economy. An international airport – a *conditio sine qua non* for tourism development – provides improved access to other regions for locally produced goods.

Another benefit is the generation of entrepreneurial activity. According to

Mathieson and Wall (1982), the extent to which the tourism sector can establish linkages with local entrepreneurs depends on:

- the types of suppliers and producers with which the industry's demands are linked;
- the capacity of local suppliers to meet these demands;
- the historical development of tourism in the destination area;
- the type of tourist development.

In terms of technical polarization one can identify backward linkages. When a number of big hotels are located in a region, there is an immediate demand for large volumes of agricultural products and different kinds of services. Local suppliers are often unable to meet this demand in quantity and quality. After a number of years, however, the imported supplies might decrease and the local supplies increase, much depending on the capacity of local suppliers to meet the new demands. Entrepreneurial activity may be further stimulated by the external economies created (Krippendorf *et al.* 1982; Vanhove 1986; De Kadt 1979; Krippendorf 1975, 1987; Mathieson and Wall 1982; Bull 1994; Frechtling 1994).

COSTS

Preliminary considerations

Each coin has two sides. Benefits have been dealt with above. However, there is no economic activity or project without costs. A distinction is made between private costs (e.g. a hotel) and external diseconomies. The costs the latter impose are called incidental costs. The sum of private costs and incidental costs is called the social cost of an activity. There is an extensive literature dealing with benefits and costs of tourism. Some authors have emphasized the cost side (De Kadt 1979; Krippendorf 1975), others have stressed the benefits (Archer 1991) and a third category pays attention to both sides of the coin (Bull 1991; Frechtling 1994; Mathieson and Wall 1982).

Often the remark is made that few studies have attempted to pay attention to economic costs of tourism in a systematic way. Mathieson and Wall assert that research has been limited largely to the measurement of the more obvious costs such as investment in facilities, promotion and advertising, transportation and other infrastructure. Most studies have failed to address the indirect costs, such as the importation of goods for tourists, inflation, the transfer of the profits, economic dependence and opportunity costs (Mathieson and Wall 1982). Nevertheless a correct assessment of benefits takes into account a number of these qualifications.

As noted above, all transfers and imports were eliminated. As a consequence, leakages are taken into account and the tourist income multiplier is lower. However, local inhabitants might change their buying behaviour due to a demonstration effect of tourists (e.g. purchase of imported products instead of local ones).

It is agreed that opportunity costs must be considered. If labour or land is used for tourism its social cost to an economy is its opportunity cost, or the lost opportunity of using it in the (presumably) next-best activity. As already noted, in most tourist countries or regions with a tourism vocation there is no full employment. The choice is seldom between industry and tourism, but between tourism and unemployment. Even when an alternative activity can be retained, a tourism region with valuable resources starts from the free raw materials as a main advantage. Is it not remarkable that many objective 1 regions of southern Europe, eligible for European Regional Development Fund support, have opted for tourism as a strategic development path?

Overdependence on tourism can be a danger. The sensitivity of tourism demand to all kinds of external factors has been underlined above. Tourism is susceptible to changes from within (e.g. price changes and changing fashions) and without (e.g. global economic trends in the generating markets, political situations, religious confrontations and energy availability).

Mass tourism and inflation

'Tourism produces inflation' is a very dangerous slogan. The relationship between tourism and inflation is more complex, temporal and local.

A high inflow of tourists during a season can provoke a rise in prices of many goods and services in the tourist region. Durand *et al.* (1994) assert that it is indisputable that in cities and tourist areas prices for products and services are, in general, higher than in cities or regions where there is little or no tourism, and that in holiday resorts prices for tourist services are higher in the peak season than for the rest of the year. This upswing of prices is presumably higher in poor regions than in richer ones. Tourists can afford to buy items at high prices. Retailers increase their prices on existing products and provide more expensive goods. This has a double consequence. First, local residents have to pay more for their goods. Second, retailers selling to tourists can afford to pay higher rents and taxes which are passed on to the consumer (Mathieson and Wall 1982).

But how far does one notice the spatial impact? Tourist demand is very often concentrated in a limited number of streets or areas. Local residents change their buying behaviour and move to other points of sale. Furthermore, tourists in general are only interested in a narrow range of goods and services such as souvenirs, sports articles, clothes, beauty products, meals and special products (e.g. chocolates and lace in Bruges).

A different aspect is the price evolution of accommodation (hotels, rented apartments) and other facilities. In the short term, supply is inelastic and an upswing of mass tourism in a region may lead to higher prices. There is not always much discipline in the tourism sector. A substantial increase in demand is followed by price increases. Regions very often forget that they are in competition with other regions. The classical movement of demand from one Mediterranean country to another is a well known phenomenon.

It is said that mass tourism makes land prices higher. Growth of tourism creates additional demand for land, and competition from potential buyers forces the price of land to rise. The local inhabitants are forced to pay more for their homes. Are the increasing land values to be considered as negative? All owners, land owners and local residents profit from the additional value. From the macroeconomic point of view the final result is a benefit. Furthermore, this effect is quite local.

All in all the impact of mass tourism on local residents should not be overemphasized. The costs are largely compensated by the benefits: more wealth, more jobs and higher land value. However, one can imagine situations – when tourism demand is very high – where inflationary tensions in tourism spill over into the economy at large and contribute to a rise in general inflation. In some countries tourism demand represents 10 per cent and more of GDP, and intersectoral linkages of tourism are intensive.

Another question is: what are the factors responsible for inflationary pressure in the tourism sector? The French authors Durand *et al.* (1994) make a distinction between demand and cost inflation in the tourism sector. First of all there is a demand inflation which results in a number of characteristics:

- seasonal demand;
- inelastic supply;
- insufficient market reaction (certain resorts or firms profit from an economic rent);
- imported inflation due to international arrivals (impact of hard currencies and increase of the money mass).

Cost inflation is a consequence of a number of factors:

- peak management;
- high taxes on some tourist products and services.

However, one cannot increase prices without considering the consumer, given the law of supply and demand which moderates prices. Many tourists changed their destination from France to Spain, and from Spain to other destinations, because of price differentials.

Incidental costs of tourism

The costs emphasized by De Kadt, Krippendorf and many other authors are summarized in an excellent way by Frechtling (1994) and covered by the term 'incidental costs' or detrimental externalities or external diseconomies.

Incidental costs, according to Frechtling, lead to quality-of-life costs and public or fiscal costs. Indeed, the local population of a region affected by external diseconomies of tourism can choose to deal with them in one of three ways:

1 they may or have to accept a lower quality of life than they enjoyed without tourists;
2 they may redress the decline in their quality of life through public expenditure for which they pay taxes;
3 they may directly impose monetary costs on tourists through taxes and fees.

In Table 4.7 Frechtling (1994) summarizes a number of important categories of incidental costs that are related to tourism import. It is not certain that a specific volume of tourists will produce costs in all categories.

Besides direct incidental costs, Frechtling (1994) distinguishes secondary

Table 4.7 Possible direct incidental costs of tourism

Life-quality costs	Fiscal costs
Traffic congestion	Highway construction, police services, public transportation, port and terminal facilities
Crime	Police services, justice system
Fire emergencies	Fire protection
Water pollution	Water supply and sewage treatment
Air pollution	Police services, public transportation
Litter	Solid waste disposal, police services
Noise pollution	Police services, zoning
Destruction of wildlife	Police services, park and recreation facilities, forestry maintenance, fish and game regulation
Destruction of scenic beauty	Park and recreation facilities, police services
Destruction of social/cultural heritage	Maintenance of museums and historic sites, police services
Disease	Hospital and other health maintenance facilities, sanitation facilities, food-service regulation
Vehicular accidents	Police services, justice system

Source: Frechtling 1994: 395

incidental costs. Additional visitors lead to new businesses or an extension of existing ones which in turn requires more employees and consequently more population. The latter imposes additional life-quality and fiscal costs on the community. Some of these costs for the additional residents are similar to those of additional visitors.

It is beyond the scope of this chapter to expand on these indirect costs which are not generated directly by tourists. One example is very typical. Tourism demand is in many regions seasonal and provides seasonal job opportunities. The region or country attracts labour that settles in and requires unemployment compensation and other income transfer programmes during off-peak seasons.

SOCIAL COST-BENEFIT ANALYSIS

A number of benefits and costs of mass tourism have been dealt with above. In practice it is easy to compare both impacts of mass tourism. Some items, especially benefits, can be assessed without too many problems. There are more difficulties with the measurement of costs. Assessing the real value of many incidental costs is particularly difficult.

So far the discussion has dealt with the notion of mass tourism, timeless and spaceless. In reality mass tourism is taking place in a country or a region and in a specific period. Very often it is related to a huge tourism project. The research stages to be faced in comparing social costs and benefits are fivefold:

1 identification of social costs and social benefit items;
2 quantification of the cost and benefit items;
3 valuation or translating the costs and benefits into monetary terms;
4 calculation of the NPV (net present value) and/or IRR (internal revenue rate);
5 risk analysis (Vanhove 1982).

This brings up the question of social cost-benefit analysis. Of the five stages above, attention can be directed to the first: how to identify the costs and benefit items?

A scheme as set out in Table 4.8 can be used, distinguishing four levels of costs and benefits:

1 project level, whatever it may be;
2 unpaid level;
3 underpayment level;
4 side effects.

The items mentioned in Table 4.8 are only examples and are not exhaustive. One should adapt the itemization to each practical case.

The unpaid level refers to (a) costs which are not taken into account by the paymaster of the project; and (b) benefits which do not accrue to the firm, organization or others who take the initiative. Only marginal costs and benefits should be considered.

The underpayment level demands more specification. In many tourist regions unemployment is very high. The assessment of the items in using unemployed resources should be at opportunity costs. The employment of unemployed persons in a project brings zero opportunity costs and reduces to a large extent investment and operation costs.

Very important is the spatial dimension of the project considered. A region does not take into account benefits which accrue to other regions; nor does it pay for costs financed by a higher level.

Some items are intangible or incommensurable. In this case the NPV or IRR of the cost-benefit analysis should be completed with a qualitative table in which all items not expressed in monetary terms are mentioned.

Table 4.8 Identification of benefits and costs in a cost-benefit analysis

Level	Costs	Benefits
Project level (e.g. 100,000 additional arrivals or a holiday village for 6,000 people)	Investment costs of accommodation maintenance costs operation costs	gross receipts from accommodation infrastructure specific taxes and fees paid by the tourists
Unpaid level	life quality costs fiscal costs not retained at the project level inflation costs	improvement of international liquidity position positive demonstration effect (e.g. better health care) enhancement of a view increased value of land increased value of cultural heritage reduced loss from migration
Underpayment level	opportunity costs of labour (negative costs)	consumers' surplus
Side effects	impact on competitive regions in a country loss of value added in other projects due to diversion of demand	impact on complementary firms value added created with subcontracting firms value added in restaurants, shops, etc.

CONCLUSIONS

Tourism is a mass phenomenon. All developed countries show very high holiday participation rates. These are an expression of mass tourism. Furthermore, a large number of secondary and short holidays can be noted. All forecasts predict an increase of demand in the next decade. Mass tourism is here to stay and has to be accepted.

The main benefits of tourism are income creation and generation of jobs. For many regions and countries it is the most important source of welfare. Although employment in tourism – staff function excluded – is of a relatively low profile for many regions or countries, mass tourism is one of a limited number of alternatives. Costs are inevitable for all activities. However, in tourism there are many direct incidental costs, most of which are difficult to assess.

How to cope with these negative impacts? The answer is not evident, but three suggestions of great significance are discussed.

First, more staggering of holidays in time, space and product would assist. Staggering in time means that each country must start with a national solution. Staggering in space will result from new tourist areas and new destinations. Long-haul holidays are only in a take-off phase. Staggering in product is related to social trends and trends in tourism demand. People are becoming less and less interested in group tourism. 'The consumer' no longer prevails; instead it is 'this consumer', who is becoming more and more interested in specialized products related to culture, hobbies, sports, health, nature, ecotourism, smaller-scale tourism, etc. (Weaver 1991). Structuring of products becomes an important mission of national, regional and local tourist organizations and of the private sector as well.

Second, the public sector should anticipate and avoid a number of costs. Tolerable numbers must be a central issue in the planning of resorts and tourist regions. Unfortunately the public sector often demonstrates a lack of understanding of the need for long-term strategy and has not always the courage to intervene. Lack of physical planning is very often the rule.

Third, a better behaved kind of tourist would be desirable: one with respect for culture, nature, population and higher moral values; one characterized by Krippendorf (1987) as an 'emancipated tourist'.

The title of this book is *Tourism, Sustainability and Growth*. Growth seems to be evident. Sustainability is not assured in all countries or all regions. The above-mentioned suggestions can be a step in the right direction. Further interesting considerations can be found in Poon (1993: Chapter 10) under the title 'Strategies for tourism destinations', which make a link between growth of tourism and sustainability. The basic starting point is new tourism defined:

> New tourism is a phenomenon of large-scale packaging of non-standardized leisure services at competitive prices to suit the demands

of tourists as well as the economic and socio-environmental needs of destinations.

(Poon 1993: 85)

It exists if the following conditions hold:

1 The holiday is flexible and can be purchased at prices that are competitive with mass-produced holidays.
2 Production of travel and tourism-related services are not dominated by scale economies alone. Tailor-made services will be produced while still taking advantage of economies of scale where they apply.
3 Production is increasingly driven by the requirements of consumers.
4 The holiday is marketed to *individuals* with different needs, incomes, time constraints and travel interests. Mass marketing is no longer the dominant paradigm.
5 The holiday is consumed on a large scale by tourists who are more experienced travellers, more educated, more destination-oriented, more independent, more flexible and more 'green'.
6 Consumers consider the environment and culture of the destinations they visit to be a key part of the holiday experience.

New tourism has consequences in different fields and for several parties: marketing, product policy, tourists, actors in the destination; and national, regional and local authorities.

Poon (1993) identifies four strategies that tourism destinations will need to implement in order to foster the development of a new and more sustainable tourism. These strategies are:

1 putting the environment first;
2 making tourism a lead sector;
3 strengthening distribution channels in the market place; and
4 building a dynamic private sector.

Only the first strategy has a direct link to this discussion. There are in Poon's approach three main principles associated with putting the environment first. They are:

1 build responsible tourism (e.g. control capacity; develop tourism with dignity and plan for the tourism sector);
2 foster a culture of conservation (e.g. develop awareness campaigns among the local population and tourists alike; encourage press and pressure groups to take appropriate actions and lead by example);
3 develop an environmental focus (e.g. solve environmental problems and export the solutions; exploit niches in ecotourism and move beyond ecotourism).

These are important principles, difficult to implement, but there is no other way to cope with mass tourism for the benefit of many regions and less developed countries in particular.

REFERENCES

Archer, B. (1991) 'The value of multipliers and their policy implication', in S. Medlik (ed.) *Managing Tourism*, London: Butterworth-Heinemann.

Archer, B. and Fletcher, J. (1990) 'Tourism: its economic importance', in M. Quest (ed.) *Horwath Book of Tourism* London: Macmillan.

——(1996) 'The economic impact of tourism in the Seychelles', *Annals of Tourism Research*, 1: 32–47.

Baum, T. (1995) 'Trends in international tourism', *Insights*, London: English Tourist Board.

Boerjan, P. and Lowyck, E. (1995) *Basisbegrippen Rekreatie En Toerisme*, Brussels: Tobos.

Bull, G. (1991) *The Economics of Travel and Tourism*, New York: Pitman.

Burkart, J. and Medlik, R. (1974) *Tourism: Past, Present and Future*, London: Heinemann.

De Brabander, G. (1992) *Toerisme en Economie*, Leuven: Grant.

De Kadt, E. (1979) *Tourism: Passport to Development*, Oxford: Oxford University Press.

Durand, H., Gouirand, P. and Spindler, J. (1994) *Economie et Politique du Tourisme*, Paris: Librairie Générale de Droit et de Jurisprudence.

Edwards, A. (1994) 'International tourism forecasts to 2005', The Economist Intelligence Unit, *Special Report no. 2454*.

Fink, C. (1970) *Der Massentourismus*, Berne: Verlag Paul Haupt.

Fletcher, J. E. and Archer, B. (1991) 'The development and application of multiplier analysis', in C. P. Cooper (ed.) *Progress in Tourism, Recreation and Hospitality Management*, London: Belhaven Press.

Frechtling, D. C. (1994) 'Assessing the impacts of travel and tourism: measuring economic benefits – measuring economic costs', in J. R. B. Ritchie and C. R. Goeldner (eds) *Travel, Tourism and Hospitality Research*, 2nd edn., New York: John Wiley.

Hudson, R. and Townsend, A. (1992) 'Tourism Employment and Policy Choices for Local Government', in P. Johnson and B. Thomas (eds) *Perspectives on tourism policy*, London: Mansell Publishing.

Krippendorf, J. (1975) *Die Landschaftsfresser*, Berne: Hallwag Verlag.

——(1987) *The Holiday Makers: Understanding the Impact of Leisure and Travel*, London: Heinemann, 1987, tr. of *Les Vacances, et Après*, Zurich: Orell, Füssli Verlag, 1984.

Krippendorf, J., Messerli, P. and Hänni, H. (eds) (1982) *Tourismus und Regionale Entwicklung*, Diesüsenhofen: Verlag Rüegger.

Mathieson, A. and Wall, G. (1982) *Tourism: Economic, Physical and Social Impacts*, London: Longman.

Mossé, R. (1973) *Tourism and the Balance of Payments*, Geneva: IUOTO.

Poon, A. (1993) *Tourism, Technology and Competitive Strategies*, Wallingford: CAB International.

The Host Consultancy (1991) *Jobs in Tourism and Leisure: A Labour Market Review*, London: English Tourist Board.

Theuns, L. (1975) 'Enkele economische aspecten van internationaal toerisme in ontwikkelingslanden', *Economie*, April.

UNCTAD (1971) *Elements of Tourism Policy in Developing Countries*, Geneva: United Nations Centre for Tourism and Development.

Vanhove, N. (1977) 'Fremdenverkehr und Zahlungsbilanz der EG-Länder und der Mittelmeerländer', in R. Regul (ed.) *Die Europaïschen Gemeinschaften und die Mittelmeerländer*, Baden-Baden: Nomos Verlagsgesellschaft.

——(1982) *Interrelations Between Benefits and Costs of Tourist Resources: An Economic Approach*, Zagreb: AIEST-Congress.

——(1986) 'Tourism and regional economic development', in J. H. P. Paelinck (ed.) *Human Behaviour in Geographical Space: Essays in Honour of Leo H. Klaassen*, Aldershot: Gower.

——(1993) 'Sociaal-ekonomische betekenis van het toerisme in Vlaanderen', in U. Claeys (ed.) *Toerisme Vlaanderen*, Leuven: Acco.

Weaver, D. B. (1991) 'Alternative to mass tourism in Dominica', *Annals of Tourism Research*, 3: 414–32.

Wood, R. C. (1992) 'Hospitality industry labour trends: British and international experience', *Tourism Management*, 3: 297–304.

WTO (1995) *Global Tourism Forecasts to the Year 2000 and Beyond*, vol. 1, Madrid: World Tourist Organisation.

5

THE CONTRIBUTION OF LIFE CYCLE ANALYSIS AND STRATEGIC PLANNING TO SUSTAINABLE TOURISM

Chris Cooper

This chapter integrates the two mutually compatible approaches of strategic planning and life cycle analysis to provide a guiding framework for the implementation of sustainable tourism at the destination. A review of both quantitative and qualitative parameters of tourism growth is provided and the implications for the management of tourism destinations outlined. In particular, the growth of demand in volume terms is placing pressures upon destinations while the changing nature of demand implies that destinations adopt sustainable principles. The process of strategic planning is described and the Tourism Area Life Cycle introduced as an explanatory framework for the adoption of sustainable principles. At each stage in the life cycle the competitive position of the destination demands a distinctive strategy and approach to management. The chapter outlines the options at each stage of the life cycle. The chapter concludes that only by considering life cycle stage and thinking strategically, can the true elements of sustainable tourism be achieved.

INTRODUCTION

The success of the tourism sector in the postwar period has resulted in an industry which is characterized not only by growth but also by rapid change. In the opening chapter of this volume, Pigram and Wahab suggest that the management of this growth and change will be critical in the 1990s and into the twenty-first century if tourism is to become a mature and acceptable sector. Here, concern for sustainability is identified as central to the management of this growth. This view is echoed by De Kadt (1992) who clearly states that sustainability has become the organizing concept for policy in the 1990s; a concept which has demonstrable advantages for the tourism sector as a whole (Bramwell and Lane 1993; Bramwell *et al.* 1996, World Tourism Organisation [WTO] 1993).

In tourism the adoption of sustainability has been evidenced by a changing perspective away from the short term to the long term, which in turn has seen the adoption of a strategic approach to both markets and destination planning/management. There is a clear synergy between the adoption of sustainable tourism principles and the disciplined, longer-term perspective provided by the strategic planning of both destinations and markets (Cooper 1995). The mutually complementary approaches of sustainability and the strategic perspective provide an organizing framework within which to manage both the growth of tourism and to ensure that tourism becomes a renewable resource at each stage of the tourist area life cycle (Butler 1980). This encourages adoption of the longer-term perspective with which to deliver control and responsibility, and thus prevent destinations exceeding capacity and the inevitable negative impacts of tourism which follow (Butler 1992).

This view is also supported by Bramwell and Lane (1993) in their view of sustainability as the idea of holistic planning and strategy formulation. In terms of the consequences of uncontrolled growth, Farrell (1992) justifies the holistic stance by defining sustainability as:

> essentially an exercise in the optimization and finer tuning of all elements and sub-elements of the development system so that in its operation, the system as a whole maintains direction and one or more of the elements does not surge or is not enhanced to the detriment and impairment of others.
>
> (Farrell 1992: 123)

Farrell is therefore advocating the constant review and adjustment of all elements of the destination formula with the adoption of strategies appropriate to the various stages of the tourist area life cycle. It is the aim of this chapter to illustrate the utility of both the strategic planning and life cycle approaches in the management of growth and the implementation of sustainable tourism development at the destination.

THE GROWTH OF TOURISM

Since 1950, destinations around the world have experienced pressures resulting from the growth of demand. In the period 1950–90, the growth of international tourism arrivals has averaged 7.2 per cent per annum, whilst in expenditure terms (excluding international transport) the annual average increase has been 12.3 per cent (WTO 1994). Of course these rates include the relatively rapid expansion of international tourism in the 1960s and 70s, fuelled by technology, affluence and a supply-side response on the part of the tourism sector towards the industrialization and standardization of the delivery of tourism. In the 1990s the rates of increase have moderated, although they are still positive despite both economic and political constraints in some regions. The evolutionary growth of international tourism since 1950 also demonstrates striking regional contrasts, in particular the rise of the East Asia/Pacific region at the expense of Europe and the Americas. These regional trends have been detailed earlier in this volume by Pigram and Wahab.

The WTO (1994) forecasts for tourism suggest a continuation of positive growth, with international arrivals increasing by around 5 per cent per annum during the remainder of the 1990s, and for domestic tourism to grow by around 4 per cent per annum. Of course domestic tourism represents a far greater volume of tourism in the world – estimated at around four times the volume of that of international tourism.

WTO (1994) states that the determinants of this growth in demand for tourism are both external to tourism and also influenced by the tourism sector itself. External to tourism are the general health of the economy, political influences and technology. WTO states that tourism tends to follow general economic trends – growing or contracting with economic activity. Superimposed on this general observation are the influences of technology and demographics.

The main influences upon tourism growth identified by the WTO are:

- socio-demographic – especially the ageing of populations in the industrialized countries;
- electronic information/communication systems;
- a more knowledgeable and demanding customer;
- a deregulating market place;
- pressure for responsible tourism development;
- a polarization between global players and 'niche' businesses; and
- constraints upon growth related to labour, capital and infrastructure.

These trends have a combined impact upon any destination considering future strategies to both manage and capitalize on the growth of tourism. Whilst the structure of the generating countries of international tourism is changing, the leading industrialized countries will continue to dominate. Here, WTO identifies the ageing of the postwar baby-boom population cohort, combined with the trend to independent and knowledgeable travellers (for example more women travelling) as important influences upon the development of products. Interwoven with and supporting these factors are the developments of information technology (such as computer reservation systems and the Internet) which support the independent traveller and hasten the flexibility of supply in the sector. Combined, these trends suggest an increasingly discerning and informed consumer who will choose tourism destinations on the basis of information and experience.

Clearly then, destinations must be in tune with this changing demand for tourism as consumers become more experienced, informed and discerning. Poon (1993) sees a clear evolution here from her concept of the 'old tourism' to a new and more flexible form of tourism, in part driven by demand, but also by a supply-side response. Poon's 'new tourists' have values which are oriented towards the environment and the ethical consumption of tourism, and have motivations which are related more to active rather than passive vacation activities and the search for escape and for the authentic. The approach of the tourism sector to market segmentation is therefore inadequate as traditionally simple demographic, geographical and purpose-of-visit categories have been used. Instead, techniques such as psycho-graphic segmentation are more appropriate to the analysis of Poon's new tourists. However, there are real problems of both measuring the defining variables of these new tourists, and of the clear identification of groups. For example, Poon (1993: 114) provides the following list of characteristics of the new tourist. They have:

- more experience
- changing lifestyles
- changing values
- more flexibility

Also, they are both more independent-minded and the products of changing demographic structures. The lifestyle and demographic variables can be quantified but the variables related to independence, experience, values and flexibility demand a more subtle and qualitative market-research approach (well developed in psycho-graphic segmentation, but which is both expensive and often beyond the reach of the expertise of tourist destinations). Indeed, for many destinations even simple measures of volume and visitor characteristics are elusive.

On the supply side, the continued increase and changes in the nature of demand will place pressures upon destinations in particular to adopt sustainable practices. This has been fuelled by the combination of media-led concern for responsible tourism, lobbying by industry groups such as the World Travel and Tourism Council, and the development of both international and also more locally based pressure groups such as Tourism Concern. These trends all point in one direction – the imperative for the development of tourism 'in a form, on a scale and to standards which are sustainable' (WTO 1994: 10).

STRATEGIC PLANNING AND LIFE CYCLE APPROACHES

In Chapter 2 of this volume, Pigram and Wahab develop the concept of a tourist area life cycle (TALC) and state that whilst the ultimate result of the cycle is decline, this can be prevented with sound planning and management. In other words, by the management of growth and change, sustainable tourism can be achieved at each stage of the cycle.

There is, however, a continuing debate relating to the value of the life cycle as an analytical technique for the marketing and planning of destinations. A number of authors have emphasized the utility of the life cycle approach for strategic planning (Morrison 1989; Onkvisit and Shaw 1986; Levitt 1965; Wind and Claycamp 1976; Hofer 1975). Yet Butler's original (1980) paper makes no such claims for the tourist area life cycle, and he saw it more as an explanatory framework for the understanding of the development of tourism destinations. However, by integrating the long-term perspective provided by the life cycle with the concept of strategic planning, it is possible to devise appropriate management and planning strategies for destinations at various stages of the life cycle. Kotler (1986) provides a clear definition of strategic planning as

> the managerial process of developing and maintaining a strategic fit between the organization's goals and capabilities and its changing marketing opportunities.
>
> (Kotler 1986: 58)

The defining characteristics of the strategic planning approach are the adoption of a long-term perspective; the development of an holistic and integrated plan which controls the process of change through the formation of goals; and a formalized decision process focused on the deployment of resources which commit the destination to a future course of action (Brownlie 1994; Weitz and Wensley 1984). Haywood (1990) views the approach as an extension of the one-dimensional marketing and planning approaches so often adopted in the short term by destinations. Strategic planning integrates these two approaches into a higher-order, formalized planning approach.

In terms of the adoption of sustainable tourism, the benefits of the strategic planning approach to the destination are clear (Cooper 1995). The process of goal setting provides a common sense of ownership and direction for the myriad stakeholders, whilst at the same time sharpening the guiding objectives of the destination. The coherence provided by this approach provides a framework for joint initiatives between the commercial and public sectors and demands the clear identification of roles and responsibilities. Finally, the approach delivers a range of performance indicators against which the destination's performance can be judged. In other words, strategic planning offers an integrated approach to the management of the destination and provides a sense of ownership for the stakeholders.

However, the adoption of sustainable principles, and indeed a longer-term strategic planning perspective by tourist destinations, is problematic. Simply put, the adoption of strategic planning at the destination is not as straightforward as in a commercial organization where responsibilities and reporting lines are well defined. In this respect, both Haywood (1990) and Pigram (1992) identify a number of implementation gaps in the adoption of strategic planning at tourist destinations. Destinations are comprised of a constantly shifting mosaic of stakeholders. Each of these groups has a different view of the role and future of tourism at the destination, and therefore the adoption of strategies becomes a political process of conflict resolution and consensus (Jamal and Getz 1995) all set within a local legislative context and where power brokers have a disproportionate influence (Farrell 1992).

In addition, the tourist sector at destinations is characterized by fragmentation and a dominance of small businesses which often trade seasonally. This has led to a lack of management expertise at destinations, a divergence of aims between the commercial and public sectors and a short-term planning horizon which in part is driven by public sector, twelve-monthly budgeting cycles, but also by the tactical operating horizon of small businesses. At the same time, the stage of the destination in the life cycle also influences the acceptability of planning and marketing. In the early stages of the life cycle for example, success often obscures the long-term view, whilst in the later stages, particularly when a destination is in decline, opposition to long-term planning exercises may be rationalized on the basis of cost. Also, the performance indicators adopted in such exercises can be controversial, as tourist volume is the traditional, and politically acceptable measure of success in many destinations. From the point of view of a sustainable tourism planning exercise, such measures are more likely to be the less tangible ones of environmental and social impacts.

THE STRATEGIC PLANNING PROCESS

There are a number of approaches to strategic planning, ranging from the evolutionary approach of Mintzberg (1994) and Ritchie's (1994) 'destination visioning' to the more formalized 'rational model' or 'planning school' approach. It is the latter approach which is detailed below.

The strategic planning process can be viewed as a series of iterative stages. The sequencing of these stages varies according to the objectives of the plan, but generally the following stages are identified in the tourism literature (Athiyaman 1995; Middleton 1988; Morrison 1989; Witt *et al.*; Ritchie 1994; Heath and Wall 1992; Kotler *et al.* 1993).

Situation analysis/environmental scanning/place audit

This stage essentially provides a set of data relating to the current situation of the destination. Here, an appraisal is made of the destination's competitive situation, often through the reconfiguration of the data into a SWOT analysis. The situation analysis considers data relating not only to the destination itself, but also to the external environment. Here, the issues identified above concerned with measuring demand for a destination can be problematic for smaller destinations.

Objectives and goals

The data base provided in the first stage allows the identification of key issues from which objectives and goals can be distilled. These objectives and goals provide a direction to the planning process. It is normal for destinations to include objectives and goals relating to both marketing and development, and for time frames and responsibilities to be attached. At this stage, parameters to be used to judge the success or otherwise in meeting the objectives will also be identified.

Strategy formulation

This stage involves the identification of a business portfolio for the destination through a process of decisions relating to which markets and products should receive more or less emphasis. This is in recognition of the fact that destinations can be thought of as an amalgam of different products appealing to different market segments at each stage of their development (see below).

Marketing, positioning and mix

This stage can be broken down into a series of smaller stages, beginning with the identification of target markets followed by the development of product positioning as a means of differentiating the destination from the competition. Finally, the position in the market is communicated through manipulation of the marketing mix with objectives and targets for each market segment and product.

Implementation and monitoring

It is this final stage where many of the problems of adopting strategic planning at tourist destinations are seen. The implementation process involves 'unbundling' the plan, assigning roles and responsibilities and putting into place a monitoring system. Finally, there is a need for measurement, feedback and control systems to ensure continued implementation of the plan.

THE INTEGRATION OF STRATEGIC PLANNING AND LIFE CYCLE ANALYSIS

The approach adopted for strategic planning will be dependent upon the destination's stage in the TALC. In particular the destination's point in the evolution of its development, combined with its competitive position, will determine the strategic options available. This is known as life cycle analysis (Jain 1985; Knowles 1996) and is a development of the product portfolio

Table 5.1 Jain's life cycle analysis matrix

Competitive position	Stages of industry maturity			
	Embryonic	Growth	Mature	Ageing
Dominant	Fast Grow	Fast Grow	Defend position	Defend position
	Start-up	Attain cost leadership	Attain cost leadership	Focus
		Renew	Renew	Renew
		Defend position	Fast Grow	Grow with industry
Strong	Start-up	Fast Grow	Attain cost leadership	Find niche
	Differentiate	Catch-up	Renew, focus	Hold niche
	Fast Grow	Attain cost leadership	Differentiate	Hang-in
		Differentiate	Grow with industry	Grow with industry
				Harvest
Favourable	Start-up	Differentiate, focus	Harvest hang-in	Retrench
	Differentiate	Catch-up	Find niche, hold niche	Turnaround
	Focus	Grow with industry	Renew, turnaround	
	Fast Grow		Differentiate, focus	
			Grow with industry	
Tenable	Start-up	Harvest, catch-up	Harvest	Divest
	Grow with industry	Hold niche, hang-in	Turnaround	Retrench
	Focus	Find niche	Find niche	
		Turnaround Focus	Retrench	
		Grow with industy		
Weak	Find niche	Turnaround	Withdraw	Withdraw
	Catch-up	Retrench	Divest	
	Grow with industry			

Source: Jain 1985

approach. Jain (1985) has developed a matrix which summarizes the key dimensions of this approach – an evolutionary element of life cycle stages from embryonic to ageing, and a market competitive position from dominant to weak. Within the matrix appropriate strategies can therefore be identified, dependent upon a destination's stage in the life cycle and competitive position (see Table 5.1).

However, whilst many destinations may know intuitively their position within the life cycle, this is more difficult to quantify. Here, Knowles (1996) identifies eight factors which can assist in identifying the life cycle stage:

- market growth rate
- growth potential
- range of product lines
- number of competitors
- distribution of market share amongst competitors
- customer loyalty
- entry barriers
- technology

Another approach is to consider instead growth indicators (Cooper 1992) such as

- rates of volume growth
- ratio of repeat to first-time visitors
- length of stay
- visitor profiles
- expenditure per head
- visit arrangement (package/independent)

For the competitive dimension on the matrix, Porter's (1980) classic analysis of competitive forces assists in the identification of a destination's competitive position:

- the entry of new competitors
- the threat of substitutes
- bargaining power of suppliers
- marketing power of suppliers
- rivalry amongst existing competitors

Quite simply, at each stage of the life cycle the mix of evolutionary and competitive forces differs, and thus strategies to deliver a sustainable tourism industry at the destination should be distinctive at each life cycle stage.

STRATEGIC OPTIONS

By adopting Jain's (1985) approach to life cycle analysis, it is possible to define the strategic options available for destinations (Table 5.1) and to outline the destination's characteristics and possible responses to be taken (see Table 5.2). By considering each life cycle stage for a destination, the following descriptions of the TALC stages and strategic responses can be produced.

Exploration and involvement

Here, the resort is visited by a small volume of explorer-type tourists who tend to shun institutionalized travel. The natural attractions, scale and culture of the resort are the main draw, but volumes are constrained by lack

Table 5.2 Implications of the tourist area life cycle

	Involvement	Development	Consolidation	Decline
Characteristics				
Visitor numbers	Low	Fast growth	Slow growth	Negative growth
Private sector profit	Negligible	Peak levels	Levelling	Declining
Cash flow	Negative	Moderate	High	Declining
Visitors	Innovative/ Allocentric	Mass market (innovators)	Mass market (followers)	Laggards/psycho- centric
Competitors	Few	Growing	Many rivals	Fewer rivals
Responses				
Strategic focus	Expand market	Market penetration	Defend share	Reposition
Marketing expenditures	Growing	High (declining %)	Falling	Consolidate
Marketing emphasis	Build awareness/ educate	Build preference/ inform	Brand loyalty	Protect loyalty/ seek new markets
Distribution	Independent	Travel Trade	Travel Trade	Travel Trade
Price	High	Lower	Low	Lowest
Product	Basic/ unstandardized	Improved/ standardized	Differentiated	Changing
Promotion	None	Personal selling/ advertising/ PR	Personal selling/ advertising/ PR/sales promotion	Personal selling/ advertising/ PR/sales promotion

Source: Doyle 1976; Cooper 1994

of access and facilities. At this stage the attraction of the resort is that it remains as yet unchanged by tourism, and contact with local people will be high. At the involvement stage, local communities have to decide upon whether they wish to encourage tourism, and if so, the type and scale of tourism they prefer. Local initiatives will begin to provide for visitors and advertise the resort, which may lead to an increased and regular volume of visitors. A tourist season and market area emerge and unforeseen pressures may be placed on the public sector to provide infrastructure and institute controls. At this point it is important to establish appropriate organizational and decision making processes for tourism to involve the local community to ensure that locally determined capacity limits are adhered to and that sustainable principles are introduced in these early stages.

Communities who decide to enter the tourism sector will find that, in the stages of exploration and involvement, growth of tourist volume is slow; gaining awareness takes time, and indeed many resorts may not progress beyond this stage. At involvement, destinations which decide to attract tourists will need to promote their products and access distribution channels in order to build a strong market position before competitors enter. Traditional marketing thinking at this stage is outlined by Kotler (1986) who identified the early stages of the cycle as where intensive growth opportunities occur, before the destination has exploited its current opportunities in products and markets. There are three major possibilities here: market penetration involves attracting more visitors from its existing markets to the resort without changing the product; market development involves seeking new markets as a source of visitors; and product development involves increased visits through developing closely related, new or improved products (Kotler 1986).

Development

By the development stage, large numbers of visitors are attracted and the organization of tourism may change as control passes out of local hands, and companies from outside the area (often national or multinational) move in to provide products and facilities. These enterprises may have differing aims and timescales from those of the local community in terms of sustainable development. It is therefore at this stage that problems can occur if local decision taking structures are weak. Control in the public sector can also be affected as regional and national planning may become necessary in part to ameliorate problems, but also to market to the international tourist-generating areas, as visitors become more dependent upon travel arrangements booked through the trade. This is a critical stage, as these facilities, and the changing nature of tourism, can alter the very nature of the resort and quality may decline through problems of over-use and deterioration of facilities.

Clearly then, it is in these early stages of the cycle that community initiatives for tourism and sustainable development strategies are most appropriate in order to guide the resort's future development. In the development stage emphasis changes to the management of market expansion through the careful matching of appropriate tourism market segments to particular products and destinations to provide a truly sustainable destination. The detailed marketing actions include building market share through increased visitor numbers and pre-empting competitors' customers, as well as relying on word-of-mouth promotion as early 'adopters' spread the word. It is important at this stage to look to diversifying distribution channels, maintaining the quality of the resort's products, continuing promotion to keep awareness high and adding new facilities. Kotler (1986) summarizes this stage as when integrative growth opportunities occur. This involves gaining control of distribution systems. This is only really applicable to destinations where computer reservations systems can be developed and used as a competitive tool; or where regional alliances of destinations/resorts can occur.

Consolidation and stagnation

In the later stages of the cycle, the rate of increase of visitors declines though total numbers are still increasing. The resort is now a fully fledged part of the tourism industry with an identifiable recreational business district. At stagnation, peak tourist volumes have now been reached and repeat visits from more conservative travellers dominate. Business use of the resort's extensive facilities is also sought, but generally major promotional and development efforts are needed to maintain the number of visits. Resorts in this stage often have environmental, social and economic problems.

As consolidation and stagnation approach, defence of market share against competitors becomes important, as does maintaining margins and cashflow by cost control and avoiding price wars (for example in the accommodation sector). By this stage the competition for visits is fierce and comes from a number of well entrenched, mature resorts; but a good strategy is to test market with new segments and to consolidate use amongst current visitors, to modify the product through improvements in quality, features and style, and to modify elements of the marketing mix (Kotler 1986). In other words, once visitor numbers stabilize, management should not await decline as inevitable but should seek to revitalize visits by seeking new markets, repositioning the resort, or looking at new uses for facilities.

Decline

Visitors are now being lost to newer resorts and a smaller geographical catchment for day trips and weekend visits is common. From the point of view of sustainable principles, the decline stage is very problematic. The destination has been adapted and designed to handle large numbers of visitors, and once they cease to visit, the very economic, social and environmental viability of the destination is threatened. Destination managers may recognize this stage and decide to 'rejuvenate' or 'relaunch' the destination by looking at new markets or developing the product. Introduction of new types of facility such as a casino, as at Scheveningen in The Netherlands (Weg 1982) and Atlantic City, USA (Stansfield 1978) is a common response. As at stagnation, a destination should seek to protect its traditional markets whilst also seeking new markets and products such as business, conference, or special-interest tourism. This helps to stabilize visitation, may combat seasonality, and reduces dependence on declining market segments.

Strategies for the decline stage are much more difficult when managers are dealing with the built fabric of tourist destinations rather than with a consumer product. Indeed, it is at this stage that the analogy of a product life cycle and the tourist area life cycle breaks down, simply because tourism is so closely woven into the very way of life of the town and supports jobs, services and carriers (Cooper 1990). Here, the longer-term view on the scale of the life cycle can provide managers with an historical perspective and prompt a new scale of strategic thinking – which will include the option of leaving the tourism industry completely.

The nature of destinations and their markets means that many of the opportunities identified above are led by product development initiatives. At each stage in the life cycle new product development is essential if sustainability is to be achieved. New product development spreads the risk of depending upon a small range of traditional destination formulae by allowing diversification. New product development therefore enhances the competitive position of a destination. Of course, if the process is flawed then the product will fail. This may be through poor assessment of demand and market needs, or through poor product design and positioning and/or inadequate communication of the benefits (Kotler 1986; Moutinho 1994). In tourism in particular, the fragmentation of supply can also act as a significant constraint on the development of new products at the destination level.

New product development has received considerable attention in the marketing literature (see Kotler 1986). Tellis and Crawford (1981) for example see new product development as closely linked to the evolution of a market. They suggest that products are in a constant state of evolution, influenced by market dynamics, managerial creativity and government intervention. Here, five basic stages can be identified:

divergence	the start of a new product type, which may be completely new or a modification of an existing product type (the European air-inclusive tour emerges in the 1950s)
development	sales increase and the product is adapted to suit better

consumer needs (in the 1960s the air-inclusive tour develops as a standardized 'package' of accommodation, flights and transfers)

differentiation   here, a successful product is differentiated to suit varying consumer interests (by the mid-1970s the air-inclusive tour introduces a range of destinations and formats and to suit differing markets)

stabilization   the product changes little, but the packaging, accessories, etc. are constantly varied (from the mid-1980s and into the 90s, the air-inclusive tour introduces new brochure formats and varies the market mix)

demise   the product fails to meet consumer expectations and may be discontinued (often forecast for the air-inclusive tour, but no sign of this in the 1990s)

The new product development literature envisages a series of sequential stages of development representing a continuing, ongoing process (Tellis and Crawford 1981: Figure 1). Of course, the development of new products in tourism must be guided by the overall strategic objectives of the destination. From this point of view, Moutinho (1994) identifies a number of possibilities:

- a completely new product offered to the market (for example development of a conference centre in the decline stage);
- repositioning of an existing product (for example the reconfiguration of a traditional outdoor recreation product into an ecotourism formula); or
- modifications of an existing product (for example the fine-tuning of a destination's accommodation/attraction mix to fit identified market needs more closely).

Again, these options must be related to Jain's matrix of life cycle stage and competitive position (1985: Table 1). For example, a destination at the later stages of its life cycle and in a weak competitive position may look to repositioning and modifications rather than launching completely new products.

Consideration of new product development highlights one of the major handicaps to the operationalization of the TALC approach to strategic options; that of the level of generalization (Haywood 1990; Cooper and Jackson 1989). At the level of the destination a TALC is often identifiable, yet in terms of new product development it is more likely that the destination will identify product types within the destinations such as business tourism, adventure tourism or ecotourism. Each of these product types will also exhibit life cycles, but these may not conform to that of their parent destination. At a lower level of generalization, individual enterprises within these product types will also exhibit life cycles. It is therefore important to be very clear-sighted in identifying the level of generalization for the destination strategy and new product development initiatives.

CONCLUSION

Kotler (1986) notes that the life cycle presents two major challenges:

> First, because all products eventually decline, the firm must develop a process to find new products to replace ageing ones (the problem of new product development). Second, the firm must understand how its products age, and adapt its marketing strategies for products as they pass through different life cycle stages (the problem of life cycle strategies).
>
> (Kotler 1986: 334)

This chapter has attempted to demonstrate the utility of adopting life cycle analysis to the concept of sustainable tourism at the level of the destination. The concepts of strategic planning and new product development have been integrated within life cycle analysis in order to provide an organizing framework for sustainable tourism. In other words, it is important to recognize life cycle stage before sustainable tourism initiatives are attempted at the destination. Whilst this may seem an obvious statement, it is more difficult to achieve in practice. To this end, the chapter has provided a means of analysing strategic options at each stage in the tourist area life cycle. Only by considering life cycle stage, and thinking strategically, can the true elements of sustainable tourism be achieved. Sustainable tourism must embrace both a long-term perspective and a holistic view of the destination; each of these is addressed by integrating life cycle analysis with strategic planning as practical frameworks within which to plan sustainable tourism at each stage of the tourist area life cycle.

REFERENCES

Athiyaman, A. (1995) 'The Interface of tourism and strategy research: an analysis', *Tourism Management*, 16 (6): 447–53.

Bramwell, W. and Lane, B. (1993) 'Sustainable tourism: an evolving global approach', *Journal of Sustainable Tourism*, 1 (1): 1–5.

Bramwell, W., Henry, I., Jackson, G., Prat, A. G., Richards, G. and van der Straaten, J. (1996) *Sustainable Tourism Management: Principles and Practice*, Tilburg: Tilburg University Press.

Brownlie, D. T. (1994) 'Strategy planning and management', in S. F. Witt and L. Moutinho (eds) *Tourism Marketing and Management Handbook*, 2nd edn., New York: Prentice Hall, 159–69.

Butler, R. (1980) 'The concept of a tourist area life cycle: implications for management of resources', *Canadian Geographer*, 24: 5–12.

——(1992) 'Alternative tourism: the thin end of the wedge', in V. L. Smith and W. R. Eadington (eds) *Tourism Alternatives*, Philadelphia: University of Pennsylvania Press, 31–46.

Cooper, C. P. (1990) 'Resorts in decline: the management response', *Tourism Management*, 11 (1): 63–67.
——(1992) 'The life cycle concept and strategic planning for coastal resorts', *Built Environment*, 18 (1): 57–66.
——(1994) 'Product life cycle', in S. F. Witt and L. Moutinho (eds) *Tourism Marketing and Management Handbook*, 2nd edn., New York: Prentice Hall, 341–45.
——(1995) 'Strategic planning for sustainable tourism: the case of the offshore islands of the UK', *Journal of Sustainable Tourism*, 3 (4): 1–19.
Cooper, C. P. and Jackson, S. (1989) 'Destination life cycle: the Isle of Man case study', *Annals of Tourism Research*, 16 (3): 377–98.
Doyle, P. (1976) 'The realities of the product life cycle', *Quarterly Review of Marketing*, Summer, 1–6.
Farrell, B. (1992) 'Tourism as an element in sustainable development: Hana, Maui', in V. L. Smith and W. R. Eadington (eds) *Tourism Alternatives*, Philadelphia: University of Pennsylvania Press, 115–32.
Haywood, K. M. (1990) 'Revising and implementing the marketing concept as it applies to tourism', *Tourism Management*, 11 (3): 195–205.
Heath, E. and Wall, G. (1992) *Marketing Tourism Destinations,*, New York: Wiley.
Hofer, C. W. (1975) 'Toward a contingency theory of business strategy', *Academy of Management Journal*, 18: 784–809.
Jain, S. C. (1985) *Marketing Planning and Strategy*, Cincinnati, OH: South Western.
Jamal, T. and Getz, D. (1995) 'Collaboration theory and community tourism planning', *Annals of Tourism Research*, 22 (1): 186–204.
Knowles, T. (1996) *Corporate Strategy for Hospitality*, Harlow: Longman.
Kotler, P. (1986) *Principles of Marketing*, 3rd edn, New York, Prentice Hall International.
Kotler, P., Haider, D. H. and Rein, I. (1993) *Marketing Places*, New York: Free Press.
Levitt, T. (1965) 'Exploit the product life cycle', *Harvard Business Review*, 43: 81–94.
Middleton, V. (1988) *Marketing in Tourism*, Oxford: Heinemann.
Mintzberg, H. (1994) *The Rise and Fall of Strategic Planning*, New York: Free Press.
Morrison, A. (1989) *Hospitality and Travel Marketing*, New York: Delmar.
Moutinho, L. (1994) 'New product development', in S. F. Witt and L. Moutinho (eds) *Tourism Marketing and Management Handbook*, 2nd edn, New York: Prentice Hall, 350–3.
Onkvisit, S. and Shaw, J. J. (1986) 'Competition and product management: can the product life cycle help?' *Business Horizons*, 29: 51–62.
Pigram, J. J. (1992) 'Alternative tourism and sustainable development' in V. L. Smith and W. R. Eadington (eds) *Tourism Alternatives*, Philadelphia: University of Pennsylvania Press, 76–87.
Poon, A. (1993) *Tourism, Technology and Competitive Strategies*, Oxford: CAB.
Porter, M. E. (1980) *Competitive Strategy*, New York: Free Press.
Ritchie, J. R. B. (1994) 'Crafting a destination vision', in J. R. B. Ritchie and C. R. Goeldner (eds) *Travel Tourism and Hospitality Research*, 2nd edn, New York: Wiley, 29–38.
Stansfield, C. A. (1978) 'Atlantic City and the resort cycle', *Annals of Tourism Research*, 5 (1): 238–51.
Tellis, G. J. and Crawford, C. M. (1981) 'An evolutionary approach to product growth theory', *Journal of Marketing*, fall, 125–34.
Weitz, B. A. and Wensley, R. (1984) *Strategic Marketing*, Boston, MA: Kent.
Weg, H. van de (1982) 'Revitalisation of traditional resorts', *Tourism Management*, 3 (2): 303–7.

Wind Y. and Claycamp H. J. (1976) 'Planning product line strategy: a matrix approach', *Journal of Marketing*, 40: 2–9.

Witt, S. F., Brooke, M. Z. and Buckley, P. J. (1991) *The Management of International Tourism*, London: Unwin Hyman.

World Tourism Organisation (1993) *Sustainable Tourism Development: Lessons for Planners*, Madrid: WTO.

——(1994) *Global Tourism Forecasts to the Year 2000 and Beyond*, Madrid: WTO.

6

SELECTIVE TOURISM GROWTH

Targeted tourism destinations

Boris Vukonic

INTRODUCTION

From a spatial point of view, tourism worldwide has developed in relatively restricted areas. Diversity and quality of tourist resources in these areas have formed a certain hierarchy between them, directing tourism flows and matching tourists' needs with the potential of various tourist areas and centres. However, generally speaking, tourism at all levels is characterized by a marked degree of concentration. There seems little doubt that many of the negative impacts attributed to tourism have been accentuated by this process of concentration, the extent of which varies from area to area and from place to place, depending on the resource characteristics, the kind of tourism involved and the type of measure used. The degree of concentration or dispersion at a particular scale is related to the overall stage of development of the country, region, resort or city. The concentration results from the interaction of a variety of supply and demand factors which may operate in similar ways at different scales. The resources which attract tourists are numerous and varied, but at each level these attractions are generally rather limited in number, distribution, degree of development and the extent to which they are known to the tourist. At almost all levels, distance plays a major role in influencing which attractions and resorts are accessible to particular markets or tourists.

Discussion of tourism area structure is further complicated by definitional problems, with a variety of coastal and mountain human settlements and centres having some tourism function. Tourism development has brought to certain areas many positive benefits, but many negative impacts as well. At certain stages of tourism development the question of further development has arisen, and in the last decade the focus has been on sustainable development. The concept of sustainable tourism development has introduced a new concept for life in tourism areas, both for tourist hosts and guests.

Searching for some new and different models which could help in achieving the goals of sustainable tourism development, it seems that one in particular may help: the model or theory of the tourism destination. The model emphasizes dispersion of tourist flows within the destination, as there are limits to intensification of tourism development. Policy makers in tourism may encourage such dispersion as a means of spreading the benefits of tourism or reducing its negative impacts.

HOW AND WHY A TOURISM DESTINATION CAME INTO BEING

For quite a long period of time in tourism terminology, the term and notion of a 'destination' have been used. However, in contemporary theory and practice of tourism, this term from the past has a new meaning. Basically, the term and notion of a 'destination', derived from the Latin noun *destinatio, onis*, has been used (and still is in all the romance languages, modified accordingly, as well as in some Anglo-Saxon countries) as a synonym for a place, that is, an end of a journey. This has been the meaning of 'tourism destination' in the past. However, tourism development has influenced certain changes in tourism terminology, so that such a place can nowadays bear different connotations. Since it is very difficult to define its attributes, it is also quite hard to come up with a precise definition of a 'tourism destination'. Up to now a great variety of terms have been used to denote this idea, but the term which has proved most suitable both through tourism theory and its practice remains 'tourism destination', because it most conveniently sums up everything one wants to express by this term in modern tourism. Future development of tourism will most likely make it the world's leading industry. Therefore a destination is expected to play a significant role through its position in the market. This paper is intended to elaborate and discuss further the above-mentioned concept.

Professional terminology develops and changes under the influence of a certain scientific discipline or a phenomenon development. This explains how reference to the end of someone's journey can change in meaning – the term itself, and also everything it physically denotes and stands for, undergoes certain changes in order to arrive at today's meaning. Eventually 'destination' became the synonym for tourism site, zone, region, country, group of countries, even a continent. Despite some objections by critics that the original meaning of the term should not be expanded, the beginnings of the development of the new meaning of this old word can be traced to works of certain American authors (Gunn 1988). The very first definitions of 'destination' point to the fact that it basically refers to a certain segment of space or the whole of it, called the 'tourist destination zone' or 'tourist destination area'. As a rule, it also has to be attractive enough to allure tourists and be adequately equipped with all the facilities to make a tourist's stay pleasant. Such a space, therefore, must be able to satisfy the great variety of its visitors' needs, which almost always are as heterogeneous as their age, their nationality and their social or professional attitudes (Hitrec 1995).

Some authors (Kaiser and Helber 1978) argue that such spaces should 'ensure primary motives which make tourists spend more days in the spot'. Nevertheless, the great majority of empirical research does not consider the question of defining tourism destinations; the interest is merely in areas with tourist attraction, or with sites that have the potential to become such places. These areas are analysed and measured through their level of attractiveness, provision of visitor satisfaction, and destination management itself. However, most frequently they are evaluated through commercial performance in the market. Despite the fact that demands on the market regard a tourism destination in very different ways, since a tourism destination is always far from a person's place of permanent residence, the analyses always emphasize the importance of elements of attraction or resources of such a destination. Some authors (Travis 1989) call them 'significant tourist factors', which can be either already used (valuated) or which have the potential to be used (latent).

Between 1972 and 1982, Claire A. Gunn introduced an idea which was further elaborated especially in the USA and Canada. The idea relied upon space analysis and classification of a tourist destination area according to the suggested structural pattern of so-called urban, radial and extended zones. The aim was to create a theory which would underpin a better appreciation of tourism potential in the international market. This was a totally different approach to formulation of a tourist space unit, which originally was the result of an attempt to lessen the obstacles to tourism development stemming from the elements of attractions and their static nature.

This process of transformation of a tourist place into a tourism destination was encouraged by the development of mass tourism, but also by the introduction and application of marketing in tourism, both in theory and practice. Mass tourism brought many changes as well as some problems concerning space organization and its safety, which prompted regional planning to determine some rules of conduct for tourists and their hosts within the boundaries of a tourist space. The aim was to quantify all necessary capacities regarding the adequate relationship between number of tourists and the size of a certain space unit. The existing definition of a tourist place, in the narrow sense of the word, was no longer adequate for precise reference to facilities for exchange of goods and merchandise in the market. One therefore needed to introduce a more appropriate space unit which could provide a satisfactory number of elements so that this space could be 'offered' and 'sold' in the market as one integral product. These ideas were elaborated and discussed in many theoretical papers (Gunn 1972, 1979, 1988, 1994; Schwarz 1976; Zolles *et al.* 1976; Tabares 1983; Jefferson and Lickorish 1988; Vukonic 1988, 1995; Travis 1989; Morrison 1989; Holloway and Plant 1992).

In order to comprehend the newly established differences between the notion of the destination or the final endpoint of a journey on the one hand

and the integral space (market) unit on the other, one has to be reminded of the past: that is, of the first days of the modern phenomenon called tourism.

In the attempt to explain relationships in tourism, theoreticians used to define the final spot of a journey (seen as a space) as a 'tourist place' or 'tourist resort'. In theory, the aim of a journey was seen as a place, or certain spot in an area. Such theories were additionally supported in practice by travel to such places and by the events and performances that were taking place in settlements for visitors. Since a special category of visitors, namely tourists, was involved, logically, the word 'tourist' was placed in front of the word 'place', so the acceptable term 'tourist place' or 'resort', came into being. 'Historically the evolution of tourism has been closely identified with the beginnings and subsequent development of resorts' (Medlik 1993: 126).

The unique characteristic of such a tourist place is the fact that it is firmly fixed in space within strictly set boundaries. It is important to mention that these settlements were not originally built for tourists; they existed long before tourists started to come. Moreover, the very historical dimension of the place made it touristically attractive.

Such development of tourism terminology has encouraged the development of tourism in certain naturally attractive sites and places which also had some other functions. Nevertheless, new visitors were attracted to them primarily because of their natural beauty, and especially by cultural and historical traces that were to be found in such sites. As the number of arrivals increased, new tourism infrastructure facilities had to be built in such places, which sometimes drastically changed their outlook, often not following the basic character of their well known historical urban core.

In countries and places where history took a somewhat different course, due to the level of development, it was impossible to consider tourism. This was the case, for example, in Eastern Europe. These countries did not experience the stage in the process of tourism development which all other countries in the rest of Europe have gone through – the well known period of organizing grand tours which is considered to be a forerunner of tourism development. Although many cities in these countries did have the same cultural potential as their neighbouring countries, tourists were not attracted.

A very good example of such tourism development is the coastal part of Croatia (the Adriatic). The above-mentioned situation, in general, influenced tourist places along this coast, which are situated at naturally appropriate and attractive locations. They were already populated, but according to a specially designed urban and architectural pattern (Opatija, Crikvenica). Nevertheless, cultural features of such already long-populated places did not influence the fact that they later became tourist places. So establishment of tourist places on the Croatian part of the Adriatic Sea goes back to the time when such places in Italy and France came into being.

Inland, i.e. in continental parts of the majority of European tourist countries, the notion of a tourist place was connected with the development of thermal baths and spas. Such places are even older than those on the coast, primarily because chronologically, people were first interested in the health issue, and only later in swimming and sunbathing by the sea. Finally, the fact is that the very first tourist places in the Mediterranean were actually winter resorts (even sanatoriums) and not summer resorts.

Such a tourist place had to have all the necessary facilities at a tourist's disposal. At that time, the majority of tourist countries had introduced laws that enforced sanctions for the 'situation on the spot'. Therefore they first regulated the outlook and the manner of building in a particular place, and later they put forward conditions for classifying any site as a 'tourist place'. The very status of a tourist place meant various benefits for that place. This is why many tourist countries have extended these special tourist rules to wider areas, especially coastal regions (France, Spain, Turkey, Portugal, etc.).

The above-mentioned idea of a tourist place used to be popular in tourism theory and practice for a long time. Actually, there was no strong reason for change since a tourist place continued to be the place for provision of tourists' satisfaction. Nevertheless, between 1937 and 1940, shortly before the Second World War, the need for some change began to appear. This was in fact merely a hint of changes that were yet to come, because this period of time was too short for any comprehensive analysis of change in tourists' behaviour. However, the change was obvious: greater movement of tourists into space influenced by, at that time, the constantly growing mobility of tourists, thanks to more sophisticated railway networks and other means of transport. In addition, people started to use cars more frequently, but still not to the extent that this could influence changes in tourist places and the way of life found there.

Among other things, demands for new localities as tourist places came with the end of the Second World War, when tourist facility constructions had been put into effect. The process of raising and building new facilities and settlements was more or less intensive, depending on the part of the world where it was taking place. The first big change happened when the number of tourists increased. This raised the question of building bigger residential and other facilities, which simply could not be achieved within the already existing urban structure of ancient towns and settlements along the coast. Some serious difficulties emerged in densely populated European areas, especially in cities and settlements which had been formed long before tourists were attracted to them, and where infrastructure, marketing, municipal and other services and facilities had already been developed. On the other hand, the same problems could not be seen in other parts of the world where the space unit was not seen as a factor of limitation. Accordingly, a tourist place, later a tourism destination, was rarely discussed in these countries. This might have been the reason for a quite frequent lack of comprehension in viewing European difficulties in this matter. Nowadays

Europe is regarded to be touristically the most visited continent in the world (59.5 per cent of world tourist arrivals in 1995) and it was therefore the need for further development which prompted the search for appropriate available space for tourists.

The approach for construction of new tourist facilities accepted the idea of the formation of so-called tourist ensembles or complexes which would be able to solve the problem, since they would be built separately from the originally existing urban settlements. Nevertheless, the problem then became the existing definition of a tourist place, because it was obvious that the new, separately built tourist settlements and facilities would now become the final destinations for many tourists. On the other hand, such groups of buildings and facilities could not be regarded as a kind of stan-dardized settlement, and neither could they be regarded as a tourist place: they lacked the basic structure and the usual features of a settlement, but even more than this – they lacked the local population. A classic example of such development is the Mediterranean area, accounting for about 30 per cent of world's tourism, and especially its touristically most developed part: the Italian, French and Spanish coasts.

This situation still did not prompt touristically developed countries to make adequate changes in defining a tourist place. Moreover, tourist agen-cies and tour operators showed relatively weak interest in 'selling' such isolated units of tourist settlements. The reason is that tourists as a rule wanted and preferred a populated, 'live' place with a tradition of living. Such a place has its own life and functions which are convenient and attrac-tive for tourists. Still, such agglomerations were built at that time, but by the next decade they were very often being located close to already popu-lated areas. This was because the logic of money and cheaper land in such isolated localities prevailed. Obviously, the greater cost of building the infrastructure network, which was always to be considered when estab-lishing such a tourist complex, was not a strong enough reason to change the philosophy and policy of the new tourism development, partly because of the fact that the state almost always had to raise the funds for it. This problem is also very typical of Europe, since in other parts of the world the state was not usually involved with its tourism development. There were different reasons for this, ranging from the lack of financial means and poverty in general (as for example in Africa and Asia) to the comprehension of tourism as an individual leisure industry (as in the USA and Canada).

In the wider coastal tourist area of the Mediterranean, such constructions have merged with the urbanistic structure of already populated places, resulting in the creation of whole strings and belts of areas called (according to the Mediterranean tradition) 'rivieras'. The idea of 'riviera' has for the first time produced a completely new approach to tourist space: tourists are now no longer interested exclusively in a tourist place seen as a discrete unit within a limited area.

The expansion of tourism capacities and facilities within a wider space has introduced a new notion: the tourist centre. This concept has meant many different things. According to the first explanation, a tourist place is a central tourist place in a wider space, that is a tourist place of the highest rank within a certain limited space. Second, a tourist centre is generally a big or a very popular tourist place. Another definition takes it to be merely the centre of a tourist place, or a tourist spot.

Among themselves, these centres differ according to the number and extent of tourism functions which they provide for a wider area. Such tourist centres meet tourists' needs both from in and around tourist places in the particular area. They actually provide tourist satisfaction for the whole area that inclines to it. Also, a centre like this usually serves as a distributional centre of goods (not only for the goods that will provide tourist satisfaction, but also for goods needed by the local population). This is why theoreticians say the area around inclines to such centres, from which the notion of tourism destination later came into being.

This was when, due to more developed environmental planning, some new notions, such as tourist zone and tourist region, were introduced. Basically, these notions rely upon urbanistic, and generally environmental and economic criteria. In short, the main idea is homogeneity, the need for the total development of a unit in space (Gunn 1972).

The introduction of certain facilities of tourist offerings into a wider space resulted in the expected dispersion of tourists. The outcome was a marketing evaluation even of those resources which, according to common economic criteria and procedures, would otherwise represent second- or even third-rate value. Along with the primary facilities of tourist offerings such as accommodation and catering, some other facilities were gradually established, especially mercantile, sports and leisure facilities, so that some residential facilities had to be built as well. All this caused some other functions of a settlement to be introduced. These were mostly placed in the centre of a wider space unit, but according to a somewhat different environmental pattern from the one followed in traditional settlements, that is, in tourist places. Finally, it all resulted in a different outlook of a settlement and a changed visual identity.

This experiment was undertaken in Spain, and later in France and Turkey, as well as in Italy and Greece to some extent, and it proved to be very efficient in many ways. In fact, Spain had decided to stimulate the introduction of various tourist, catering, leisure, sports and marketing facilities beyond those developed spots (mostly along the coast and close to water) in order to keep the local population out of touristically developed areas. In this way the local population was spared social impacts and their living standard was raised. This new philosophy, of leading tourists inland from their place of stay, rather than encouraging the local population inland to come and stay permanently in tourist places, is one of the factors that has helped give birth to the notion of tourism destination. This same philosophy has also been applied to the numerous Greek islands.

WHAT IS THE MEANING OF 'TOURISM DESTINATION'?

Possibly the best, and also the simplest answer to this question was given by Gunn, who in *Tourism Planning: Basics, Concepts, Cases*, calls a tourism destination a 'travel market area' (Gunn 1994: 107). This is the very essence of the desired integral change: to create a complete tourist unit within an area which will be able to exist independently and efficiently in the international tourism market according to the principles of marketing and the policy of the tourist product. Obviously, new planning methods are needed to protect the indigenous qualities of communities and their surrounding areas (Gunn 1994). It is more than obvious that in the future development of world tourism, certain localities and tourist places will not be accepted as 'travel market units'. They will be seen exclusively as parts of a wider area, of a tourism destination. This means a big change in theoretical discussions, practical comprehension, and especially the management and marketing of tourism destinations, both in Europe and in other parts of the world where tourism development has just begun.

In tourism theory and practice, there is a difference between previous and current ideas of what constitutes a tourism destination. The difference can be clearly seen if one consults Medlik's new (1993) *Dictionary of Travel, Tourism and Hospitality*. It is interesting that nowadays Medlik defines 'tourism destination' in only one way, that is:

> Countries, regions, towns or other areas visited by tourists. Throughout the year their amenities serve their resident and working populations, but at some or all times of the year they also have temporary users – tourists. How important any geographical unit is as a tourism destination, is determined by three prime factors: attractions, amenities and accessibility, which are sometimes called tourism qualities of the destination.
>
> (Medlik 1993: 148)

Nevertheless, there is another term, 'resort', which can be elaborated in different ways as 'summer (winter) resort'. Such interpretation, used informally in everyday life, cannot be used in tourism terminology. Even Medlik himself states that this is in fact a place to which people go for holidays (vacations) and recreation. Thus, depending on the way of spending a holiday, there are holiday (vacation) and health resorts and as far as holiday location is concerned, there can be inland or coastal/seaside resorts. Medlik further states that this term is used for every place visited by a great number of visitors, 'and capital cities tend to be the largest and most prosperous resorts in their countries, especially for international tourists' (Medlik 1993: 126).

From the above-mentioned definition another interesting conclusion can be drawn. The notion of tourism destination, unlike the term 'tourist place', does not have strict boundaries in space. Moreover, it does not observe the same criteria upon which such boundaries could be established. Therefore, a whole country, a particular region, a certain tourist place, and even smaller locations like traffic terminals (airport, harbour, railway- or bus station) can be called a tourism destination. It is obvious that this term is used to denote the final endpoint of a journey, with no pretensions to other connotations for a completely different need. In fact this term's wide range of usage still causes difficulties in understanding the theoretical viewpoint that the notion of tourism destination is actually totally opposed to that of tourist place.

The Medlik definition, as well as others, leads to the conclusion that the term used in the contemporary theory and practice of tourism, 'tourism destination', actually corresponds to the notion and term, 'resort'. This is certainly the case in the English language. The new term 'tourism destination', with its contemporary connotations, has now entered most languages and has begun to be widely used, primarily in tourist agencies' and tour operators' practice, and subsequently by tourists themselves.

As already stated, the idea is not to consider only a certain point within an area (place, settlement) to be the final spot of a journey; the wider space around this central area (tourism destination) should also be included. The reason for this is very clear: more space means greater possibilities for provision of tourist satisfaction. This is why the USA and the Caribbean use the term 'resort' for 'a holiday (vacation) hotel providing extensive entertainment and recreation facilities' (Medlik 1993: 126). In French literature on the subject, this positive change towards more complex tourism space units begot the new term 'station', which with the addition of 'd'hiver' or 'd'été' pointed more precisely to the dominant marketing strategy of a destination.

Why is this important and how does it matter in the context of the growth and sustainability of tourism?

In order to be able to answer this question, we have to define the notion of 'tourism destination'. No doubt tourism destinations are going to make a significant impact on future world tourism development; furthermore, they will ensure this growth and development. This notion stands for a wider area, a functional unit which builds its tourist identity on the concept of cumulative attractions. The attractions provide specific experiences which, supported by the additional tourism infrastructure, together create a popular tourist area. Such a definition includes all the important components for viewing the tourism destination as a fundamental unit of tourism development and of its position in the market. It must be added that it is possible to discuss sustainability of tourism only if it includes concern for the benefit of a particular destination and its inhabitants. So it is of great significance that a tourism destination acts as an integral and functional unit in which its particular components (such as tourist places, localities, zones, etc.) can have their own specific offering, and even grow and develop independently. Nevertheless, when it comes to their presentation and their position in the market, all these components must unite with other places and localities in a wider area, so that together these represent a saleable product. Regardless of the attractiveness and the capacity of their tourist offering, such areas can be called 'tourism destinations' only if a great number of tourists is attracted to them.

The advantages of forming a tourism destination with regard to the concept of sustainability and growth of world tourism are:

- better use of the space assigned to tourism in general
- the possibility of economic evaluation of low-quality tourism resources
- the opportunity to provide tourists with a more complex service, since a wider area logically means a greater number of various tourist activities
- better conditions for creating a recognizable tourism identity and image
- better terms of presentation and promotion of a space unit in the tourism market
- a guarantee that tourists will enjoy a fulfilled stay in this space unit, which happens to be the most crucial criterion for a potential tourist in choosing a place to visit and stay

From the economic point of view, this new idea of a tourism destination includes more diverse parts of an area in the direct marketing process, which also means that less amounts of key attractive space are involved. Such situations are of special interest for land purchase policy because in this way prices rise (due to the possibility of creating various facilities for tourism development) and therefore individuals (usually local residents) and the community (country, state) are able to gain more profit. Furthermore, in this way a tourism destination is not characterized through a natural rarity, a cultural monument or some extraordinary manifestation (cultural, artistic, sports, marketing, gastronomic, etc.). These are merely the basic factors and elements upon which the complete tourist offering, tourism development and marketing image of a tourism destination are being built.

Obviously, in such a context, use of the notion 'tourism destination' for a whole country is somewhat questionable, but might be a possibility. This raises the issue of perception, the question being: how much does a particular 'spot in space' of a tourist journey 'enlarge' or 'minimize' the relationship between a tourist's residence and the potential tourism destination? For example, in so called long-haul journeys (*Fernreisen* is the German expression), tourists allow a wider area, a wider space unit, to be their journey's final focus. The case of the Japanese and Americans, who come to Europe and visit only one or a small number of tourism destinations, is widely known. To them, a tourism destination equals a whole country, and a country equals a space unit of integrated tourism potential. Moreover, Italy, for instance, is being perceived as one particular destination of European travel, France as another one, and Croatia as a third. It is quite obvious that the criteria have been formed on the basis of the earlier perception of tourism resources within a certain limited space. As soon as a tourist comes 'to the very spot', this perception changes, space criteria become narrower, so that in this example tourists who come to Italy shift the notion of 'tourism destination' to the Ligurian coast, Corsica, Florence, or even Venice.

TARGETED TOURISM DESTINATIONS

Theoreticians have tried to classify tourist destinations according to various criteria in order to, among other things, discern potential visitors more easily, namely tourists interested in one particular destination. The basic criteria in this matter were: the principle of space homogeneity or heterogeneity of a destination, meaning the level and content of a destination's attractiveness, and the number of visitors to a destination.

The principle of homogeneity and heterogeneity relies upon characteristics of fundamental components of a destination's tourism resource base. If the structure of the resource revolves around only one element, the destination is relatively homogeneous. If a destination is made up of more characteristically different but nevertheless complementary basic parts, the destination can be said to be heterogeneous or polarized. From the marketing point of view, the so-called polarized destinations are surely in a much better position, because they expand the potential marketing segments addressed by a destination. This thesis becomes more obvious with the growth of tourist demand in the market.

The level and content of attractiveness are very common criteria according to which tourism destinations are divided, since they have much to do with the notion of tourists' needs and motivations. Therefore, the concept of so-called cumulative attractions was developed. Its basic idea is that a certain tourism destination with strong and recognizable attractions within an area, or offered during a journey (presented as parts of a unique and logical system) is bound to be much more successful in the market than if the attractions were presented in a scattered manner, without any logical order or system. At this point one can raise the question of the influence which destinations exercise on each other directly or by means of feedback, seen in the light of strengthening and weakening their power of attraction and their position in the market (Hitrec 1995: 46).

Criteria for division of tourism destinations were adopted by the World Tourism Organisation (WTO) for its own statistical needs. WTO recognizes distance destinations (the most distant area from a domicile area), main destination (an area where most of the time is spent) and motivating destination (an area a visitor regards as the primary focus of a journey). Within this context even a certain hierarchy in space, where the notion of a tourism destination is used, can be noted. Therefore, in theory it is possible to refer to a central and peripheral destination, and such a division enables the creation of a constructive attitude towards the protection of attractive resources, as well as helping to realize sustainable tourism development. For example, allocation of a higher rank provides certain tourism functions for a wider area, and becomes a kind of an 'engine' promoting development and welfare for other underdeveloped zones and places (Hitrec 1995).

Since these divisions embody strong arguments for the development of a policy for tourism destinations, it is important to have access to sound information. This is possible only through market research aimed at obtaining as great a variety of different data on tourism destinations as possible. The results can be used to intervene in the structure and quality of tourism destinations' offerings, whereas both may have to be adjusted to the needs and demands of potential tourism customers. For example, research has shown that the number of journeys aimed at visiting more destinations is growing. Obviously, this means that tourist journeys are more complex than just a relationship between the outgoing and incoming point. Several models of a destination are now available, which will be of great value, especially in future tourism development (Hitrec 1993: 46). These space models are:

- journeys to one destination
- round trips with one central destination
- domicile at one destination which serves as a starting point for visiting other destinations within the area of the primary destination
- round trip within a region, visiting many destinations in the region
- 'chain journey', i.e. a typical itinerary visiting many destinations in a region

Both placing a tourism destination on the market and marketing it are closely connected to another two notions: 'life cycle' and image of a tourism destination. Both concepts rely on marketing and offer a strong basis for competition in which tourism destinations get involved in the international tourism market. The two notions must be understood in order to be able to discuss the marketing position of a destination and to proceed with business actions to influence potential segments of the demand. It is interesting that some recent research in tourism has shown that the notions of 'resort' or 'destination' bear a certain image, but always within the context of a particular wider space. Research conducted by the Marriott Corporation has shown, for instance, that in the USA the notion of 'resort' is mostly connected to places of vacation for elderly people and the upper class (Walker 1988). Contrary to this example, in the Caribbean, along with this notion, if the word 'Club Med' is added, it denotes a recreational centre with emphasis on activities for the young (Chon 1995).

The contemporary understanding of a tourism destination from the marketing viewpoint will, in future tourism development, rely upon the concept of 'life cycle'. Marketing has developed this idea strictly for the product, or to be more exact, for its conduct in the market. As it happens, marketing regards a product as any living organism which during its life passes through a series of certain phases. Each phase has its own particular characteristics: a product has to be introduced to the market; the market discovers it; it gradually grows, develops and reaches maturity. From this moment the product (sooner or later) grows old and disappears (fails) in the market. Such a concept could be used for almost any tourism destination. The most characteristic examples are tourist areas along the French coast which originally had a winter climate orientation, but in the 1920s gained all the characteristics of a summer resort. In the same manner, numerous European health resorts have undergone similar change, which has also happened to winter resorts in the Alps and the Haute-Savoie. They had to completely change the structure of their offerings, as well as their outer appearance in order to meet the needs of potential ski-tourists, and thus survive. This can be explained by the well known notion in tourism theory and practice of the constant revitalization of tourism destinations. In outlining and creating new tourism destinations, one will have to bear in mind such past experiences, because they correspond to the potential tourists' comprehension of a destination.

Within this context can be viewed the idea of the image of a tourism destination. Every tourism destination has its own respectability and credibility in the market, through which it is recognized. Of course, this image depends on a type of destination and its space boundaries. Depending on an image of a destination, the market places it either into the central or peripheral group of tourism destinations. This influences the number of arrivals at a destination, especially during seasons, as well as the complete marketing policy of each destination as a whole seen as a tourist product. The market creates an image of a destination on the basis of its own set criteria, so that a destination, by undertaking various actions in the market, cannot perform any changes in its own image. Nevertheless, it still can influence the forming of its tourism identity in a positive way, according to desired changes of its image in the market. This is not a rapid process, and destinations expecting easy and quick changes to their image are unlikely to succeed.

CONCLUSION

Contemporary tourism theory and practice, especially practice, accepts the notion of a tourism destination as an integrated tourism product. The market regards a tourism destination as a 'marketing unit', because by its cumulative attractions and tourist content, a destination better answers the needs of a potential segment of demand in tourism than any other known space unit. A tourism destination is the key to a tourist's experience. This notion also makes the business policy of a tourism destination much easier, as well as its creation and placing in the market, since the basic elements of marketing conception and strategy can always be applied. In this way, the necessary conditions for sustainable tourism growth are ensured. This is therefore the fundamental issue: in the case of a tourism destination, the application of marketing is the same as for any other kind of product. In other words, the basic ideas and strategies of marketing are totally applicable in the case of a tourism destination, and they will become indispensable in future world tourism development. Obviously, the very specific nature of a tourism destination demands that general marketing concepts be not literally and uncritically transferred to a tourism destination. They should be adjusted to the specific nature of the tourism market as well as to the specific features of each and every tourism destination, since this is the condition for a sustainable tourism development and growth anywhere in the world.

REFERENCES

Chon, K. S. (1995) 'Tomorrow's resort industry: challenges and new opportunities', *Turizam*, 43 (9–10): 159–63.

Gunn, C. A. (1972) *Vacationscape: Designing Tourist Regions*, Austin: University of Texas.

——(1994) *Tourism Planning: Basics, Concepts, Cases*, Washington: Taylor & Francis.

Hitrec, T. (1995) 'Tourism destination: meaning, development, concept', *Turizam*, 43 (3–4): 43–51.

Holloway, J. C. and Plant, R. V. (1992) *Marketing for Tourism*, London: Pitman.

Jefferson, A. and Lickorish, L. (1988) *Marketing Tourism: A Practical Guide*, London: Longman.

Kaiser, C. and Helber L. (1978) *Tourism: Planning and Development*, Boston: CBI Publishing.

Medlik, S. (1993) *Dictionary of Travel, Tourism and Hospitality*, Oxford: Heinemann.

Morrison, A. M. (1989) *Hospitality and Travel Marketing*, Albany NY: Delmar Publishers.

Schwarz, J. J. (1976) *Dynamique du Tourisme et Marketing*, Aix-en-Provence: Centre des Hautes Etudes Touristiques.

Tabares, F. C. (1983) *Producto Turistico: Bases Estadisticas y de Muestreo para su Diseno*, Mexico City: Editorial Trillas.

Travis, A. S. (1989) 'Tourism destination area development: from theory into practice', in S. F. Witt and L. Moutinho (eds) *Tourism Marketing Handbook*, New York: Prentice Hall.

Vukonic, B. (1995) 'Tourism destination: meaning and explanation of the term', *Turizam*, 43 (3–4): 66–71.

Walker, T. (1988) 'Resorts: fun-in-the-sun, plus', *Atlantic Journal*, 2 June, G1.

World Tourism Organisation (1995) *Concepts, Definitions and Classifications for Tourism Statistics: A Technical Manual*, Madrid: WTO

Zolles, H., Ferner, F. K. and Muller, R. (1976) *Marketingpraxis für den Fremdenverkehr*, Vienna: Osterreichischer Wirtschaftsverlag.

7

MODELLING TOURISM DEVELOPMENT

Evolution, growth and decline

Richard Butler

INTRODUCTION

One of the major characteristics of tourism is its dynamic nature. Like any other complex human activity, tourism is made up of a large number of elements and processes, all of which are capable of considerable change over time, in some cases at a very rapid rate. This dynamism is reinforced by marketing, for tourism is extremely competitive, and destinations have to compete aggressively for customers against other destinations across the globe. The competitive nature of tourism has increased significantly in recent decades, with changes in transportation in particular allowing potential tourists to choose from an ever-widening range of destinations and attractions. Inevitably, therefore, given the present-day western emphasis on newness and uniqueness, many destinations are deliberately changing in anticipation of, or to reflect changes in customer preferences. Thus, very few destinations remain unchanged for very long, with the limited exception of a few locations which have managed to preserve the original nature of the destination and now market that as a nostalgic or unique experience.

Such a pattern of induced and at times possibly premature change runs both counter to and in sympathy with the principles of sustainable development. One of those principles, of course, is the long-term viewpoint and commitment, and the encouragement of the dynamic element in tourism destinations can be viewed as being contrary to that ideal. Changing the nature and physical face of a destination, as well as the overall type of experience offered, may and often does involve the destination departing more and more from its original characteristics, and quite probably moving further and further from sustainability. On the other hand, keeping a destination attractive to tourists, and changing its characteristics when deemed necessary, can be argued to be one of the ways of ensuring sustainability of tourism in that destination, i.e. continuing to make the destination capable of attracting tourists over the long term. This is only one of the paradoxes involved in the development of tourism destinations, but one particularly relevant in the context of sustainable development.

This chapter focuses on the patterns and processes of tourism development with particular reference to the concept of sustainable development. While other writers in this volume have already discussed the term and some of its definitions, it is perhaps appropriate to preface further remarks with a brief review of the implications of the concept for tourism development. The literature of sustainable development has become immense over the decade since the term was introduced (WCED 1987), with only a small portion of that discussion relating directly to tourism, and an even smaller proportion actually discussing tourism. As Wall (1996) has pointed out, tourism is not mentioned directly in *Our Common Future*, and thus all of the subsequent discussion is inferring what may apply to tourism on the basis of comments made about other phenomena. We can, however, draw out some common elements relating to development in the context of tourism which would seem to be in agreement with the general principles of sustainable development. The first is that there is development involved, and that by implication this will cause change in the area involved. Related to this are the nature, scale and rate of that development, as, particularly in the case of tourism, they are of great significance in affecting that change and its acceptability. A second principle is the significance of the long-term view. If nothing else, sustainable development of any form has to imply consideration of the future, and, most would agree, the long-term rather than the immediate future.

Another fundamental characteristic implied is that of management. If development is to attain certain goals without having certain undesired impacts, then management in some manner and at some level is inevitable. Humankind has rarely proved capable of self control and giving great consideration to the future, particularly not to the long-term future, unless under the control of a despot or an all-powerful state. Even then, emphasis is often on short-term gain. To achieve sustainability, therefore, development has to be managed and controlled in order to ensure that it does not become incompatible with current and future needs. In a highly privatized and fragmented industry like tourism, this is extremely difficult to ensure. A fourth principle is the recognition that even sustainable development has impacts and effects upon most, if not all elements of the environment, physical and human, and that directing, minimizing and avoiding these impacts has to be an integral part of the development.

The major problems arise in moving from an acceptance of the above principles to their implementation in a tourism context. They represent a fundamental change in the way western society has traditionally approached and undertaken development, and as tourism for the last two centuries has been a predominantly western concept (Hartman 1986), most of the current tourism development has been by or for western cultures and their desired forms of tourist activity.

PATTERNS AND PROCESSES OF DEVELOPMENT

Tourism and the geographic environment

Although tourism has been in existence for centuries, even millennia, in very basic forms (Towner 1996), the actual development process of the phenomenon has not been paid great attention by researchers until relatively recently. While early works on tourism comment in a narrative fashion on the development of resorts in particular (Ogilvie 1933; Gilbert 1939), as might have been expected from writing of that period, analysis or modelling of the process was rarely included. In the period of explosive initial growth of tourism following economic recovery from the Second World War, some researchers did comment on both the nature of tourism development and its potential effects (Clawson 1959; Wolfe 1952). Clawson was almost alone in raising concerns about the likely effect of the rapid growth in outdoor recreation (and by implication, tourism) in the United States as the real effect of the automobile began to be felt, and pressures on resources began to increase. The nature of this increase and the changes in the character of destination areas were succinctly recognized by Wolfe (1952), who aptly used the phrase 'divorced from the geographic environment' to describe how a specific traditional cottage resort, based on a magnificent sand beach, had gradually become converted into a 'honky-tonk' amusement-dominated centre, for which the beach had become a mere backdrop. This process of conversion from inherent and generally natural attractions to created and 'artificial' features is one which underlies much tourism development, particularly, but not exclusively in the context of mass tourism destinations.

The reasons behind this process are inevitably complex and multi-faceted, and cannot be discussed in detail here. They undoubtedly reflect society's fascination with technology, with an apparent reduction in patience and a demand for instant gratification, along with an increased desire for perfection and satisfaction in leisure that cannot be obtained in work and regular living. The rise of consumerism in what is called the postmodern society, and an inability to deal with boredom or lack of activity, are reflected in a passion for entertainment rather than a desire to entertain oneself. Many of these social features extend far beyond tourism into the very fabric of daily living, but it would be surprising if they did not also manifest themselves in tourism. We may note as part of this process a broadening of the range of options of tourism destinations, with developments occurring at both ends of a spectrum simultaneously, from extremely basic and primitive facilities offered under the rubric of ecotourism and adventure tourism, to the ultimate fantasies of architecture and stimulation found at Las Vegas or the numerous theme parks.

This is not to impute value judgements with respect to which is the best or 'true' tourism, although, as will be discussed later, there are implications

111

in terms of which type of development is more sustainable. Rather it is a commentary on changing lifestyles and preferences. As noted earlier, tourism is dynamic, and it would be peculiar if destinations did not change as tourism itself changed. What is of particular significance in the context of tourism and sustainability is whether the changes are necessary, are in sympathy with the principles of sustainable development, and whether they are in fact desired by tourists, or are being undertaken inappropriately without due consideration of long-term effects and implications. Given that tourism is not homogeneous, and tourists not only represent almost all elements of human society, but also change their preferences and needs through their life cycle (Pearce 1993), the situation becomes more complex still. An additional issue is that once such development and change has occurred, it is next to impossible to have the original conditions restored, and thus the development process becomes unidirectional and irreversible.

Modelling the process

There is a growing literature on tourism development, ranging from books (e.g. De Kadt 1973; Pearce 1989; Nelson *et al.* 1993) to a large number of articles, but relatively few provide specific models of the development process of tourist destinations. In the context of sustainability and development, the focus of models should surely be upon providing a means of assessing the appropriateness of the development within the principles of sustainable development outlined above. What is necessary is not the forecasting of tourist numbers, which is rarely forecasting but more often marketing and predictive planning, but rather description and explanation of the development process. Understanding the nature of growth and the ways it changes in tourist destination areas is of major importance, not only to those involved in the tourism industry, but also to those in the public sector who have to provide much of the associated infrastructure and are responsible for maintaining environmental quality and safety, and to those who reside in tourist destinations and have to live with tourism on a full-time basis. It is perhaps less directly important to tourists, who so far have the option of taking their vacations in a different location if they dislike the direction and scale in which development has gone or is going. However, as noted below, this option may not always continue to be available.

A major theme pertaining to the development of tourist destinations which has been widely discussed in the tourism literature is that of the evolution of destinations. The use of the term 'evolution' in this context was introduced by the author (1980), although the concept itself had already attained some acceptance. It was used to imply organic change over time, generally but not exclusively in one direction, incorporating existing development, rather than a revolutionary pattern of development whereby previous developments are removed or drastically changed very suddenly.

One of the earliest proponents of the evolutionary approach to tourism destination development was the German geographer Christaller, who proposed that tourist destinations evolved from discovery to popularity to decline (1963). His ideas were based heavily upon destinations in the Mediterranean, and influenced strongly by developments in the immediate postwar decades. He provided little empirical data for the proposal and his paper was not widely circulated.

A much more influential paper was that by Plog (1974) which has appeared in several publications, and has since been widely quoted. Although it too has been criticized (Pearce 1993) for being overly simplistic and based on limited research and deductions, Plog's conclusions in particular received considerable attention. He placed various destinations on a spectrum of types of tourists, suggesting that the market for destinations would change over time, both in terms of the types of tourists and in terms of potential numbers of visitors. Plog summarized the implications of his discussion in rather dramatic fashion:

> We can visualize a destination moving across a spectrum, however gradually or slowly, but far too inexorably, toward the potential of its own demise. Destination areas carry with them the potential seeds of their own destruction, and lose their qualities which originally attracted tourists.
>
> (Plog 1974: 58)

In this sense Plog is echoing the ideas expounded by Wolfe (1952) on the 'divorce from the geographic environment', namely the way that destination areas change and frequently, (inevitably in Plog's view) obliterate or change overwhelmingly the inherent features which first made them attractive to visitors. Martin and Uysal (1990: 327) suggest that 'the move of destinations across this spectrum has become known as the tourist area life cycle'. In fact, however, the specific use of the term 'cycle' in the tourism destination context appears to have been first used by Stansfield (1978), who described the pattern of growth, decline and potential rejuvenation in Atlantic City, in a paper entitled 'Atlantic City and the resort cycle'. Also in the mid-1970s, Noronho, in a review of the sociological literature on tourism, suggested that tourism developed in three stages, which he characterized as discovery, local response and institutionalization (1974).

The reason for a significant interest in the way tourism destinations developed and changed at this time can be related to the belated interest in the impacts and effects of tourism on destination areas. Prior to the 1970s little tourism research or writing had focused on any of the negative or problematical aspects of tourism (Butler 1989). A series of publications from 1973 onwards drew attention clearly to some of the problems which could arise as a result of tourism development, and some of the unexpected and undesired changes which came about (Bryden 1973; De Kadt 1973; Smith

1977; Young 1973). The appearance of *Annals of Tourism Research* and its initial focus on the sociocultural effects of tourism development further increased attention in this area, and Mathieson and Wall's review of the impacts of tourism (1982) provided a conceptual framework for the analysis of impacts. The process of development and change of destinations did not receive similar attention, however, as the focus tended to remain on the impacts and their causes rather than changes in the environments in which they were taking place.

A very few other models did appear in this period, however, which dealt with that process. One, quoted and discussed in Pearce (1989) in some detail and more briefly elsewhere, is that by Miossec (1977), who theorized that the development process took an evolutionary form (although not using that term) and depicted the development of infrastructure and facilities in destination areas, including the coalescing of centres and the expansion of transportation networks. Lundgren (1973, 1983) echoed similar ideas in his review of the patterns of tourist development in the Laurentian area of Canada and elsewhere, although his emphasis was squarely on the transportation element. An evolutionary approach is also implicit in Lundgren's models and theories, where each spatial expansion of development is based on the infrastructure and facilities existing at the time, which are in turn superseded, and become either redundant or converted.

A widely quoted model dealing with the development of tourist destinations is that by the author (1980), which as Prosser notes 'formalizes a notion that has been traced back through the tourism literature of more than thirty years' (1995: 3). Prosser suggests three main factors have contributed to what he terms the 'enduring attraction of the destination life cycle concept' (1995: 4). One is the fact that the model provides a relatively simple conceptual framework in an area marked by the absence of theoretical approaches. A second is that the descriptive power at an overview level is inherently appealing to researchers; and a third factor suggests that research in a wide range of settings has provided qualified empirical support for the model.

The destination life cycle model

The original destination life cycle model has been described and discussed widely in the literature and does not need much discussion here. At the time that it was proposed, it was seen as a conceptual framework to illustrate and describe the process which tourist destinations undergo. It was influenced by the author's experience in Western Europe in particular, and acknowledged a debt to earlier writers including Christaller, Plog and Stansfield. The identification of seven stages in the process of development (exploration, involvement, development, consolidation, stagnation, decline and rejuvenation) was intended to clarify the evolutionary nature of development and the associated changes in a variety of elements which went with that development. It was not envisaged as a predictive model in a statistical sense, but like any general conceptual model, proposed a common pattern and process for a specific phenomenon, in this case the development of tourism at a destination. The relatively brief nature of the original work did not allow expansive discussion of some of the ideas discussed, and inevitably subsequent researchers have taken somewhat liberal interpretations of what was originally claimed and proposed.

Comprehensive reviews of the model have been made by Cooper (1994) and Prosser (1995) in particular, and it has been applied in a wide variety of situations over the intervening fifteen years (Hovinen 1982; Oglethorpe 1984; Meyer-Arendt 1985; Cooper and Jackson 1989; Weaver 1988, 1990; Choy 1992; Getz 1992; Smith 1992; Williams 1993). It has also been used in a variety of other situations examining different aspects of tourism development such as second homes (Strapp 1988), oligopoly practices (Debbage 1990), entrepreneurship (Din 1992), informal and formal sectors of local economies (Kermath and Thomas 1992), other developments (Di Benedetto and Bojanic 1993), and the relationships between national organizations and transnational companies (Iaonnides 1992). Perhaps not surprisingly, the model has been subjected to critical review and criticism (e.g. Haywood 1986, 1992; Choy 1992) and to calls for modification and alternative or additional stages (e.g. Strapp 1988; Weaver 1988; Agarwal 1994). Prosser notes, however, that 'Despite the large number of studies undertaken utilizing the conceptual model . . . and the criticism it has received, the original model survives largely intact' and goes on,

> It is unrealistic to expect unqualified explanatory power from what remains a relatively simple conceptual framework, or for the framework to account for the diverse range of factors which may influence tourism development in widely different social, ecological, economic, political and technological settings.
>
> (Prosser 1995: 9)

Implications for sustainability

It would appear then, that the principal of change in tourist destinations by an evolutionary rather than a revolutionary process has been generally accepted by many researchers and planners. The original life cycle model has been utilized in many publications and by many agencies. What then are the implications of the concept for modelling the development of tourism destinations in the context of sustainable development? The author has argued that the original paper could be seen as a call for the adoption of the principles of sustainable development long before that term was introduced (Butler 1989). One of the key aspects of the original model was that in an unchecked situation, development would continue until such time as it exceeded the innate capacity of the destination to absorb tourism and its associated development. After this point, perhaps before, problems would emerge, which if not addressed satisfactorily would result in a subsequent decline in visitation.

The relationship between carrying capacity and the tourism life cycle has been explored in some detail by Martin and Uysal (1990), who argue that 'It is impossible to determine tourist carrying capacity outside of the context of the position of the destination areas in the life cycle. The interrelationship of the two concepts is dynamic, with the idea of change implicit in both concepts' (1990: 329). It seems clear that unlimited growth in tourism in any destination is impossible and that ultimately growth rates must decline in every location (Wolfe 1966). What is of prime concern, however, is the point at which such growth should decline or cease, and also the rate at which it should be allowed. Carrying capacity in the tourism concept is a complex and multifaceted feature (Butler 1996) and researchers have long given up the search for a single number to represent the maximum number of visitors who should be allowed to be present, or the maximum number of hotels beds to be developed at a destination. This fact, however, should not be used to obscure the fact that there are, and should be recognized to be, limits to development.

Capacity takes a variety of forms: one common breakdown is into social, environmental and economic elements (Johnson and Thomas 1994). What was argued in the 1980 model was that tourist numbers and levels of development do not have to exceed all elements of capacity to have a deleterious and possibly significant effect upon the attractiveness of a destination. If the level of development, however measured, exceeds any of the elements of capacity, problems are almost inevitably going to appear. Thus it is not necessary to identify a specific upper level of development for all elements of capacity at a destination, but rather for the most sensitive ones, for they are those which will be exceeded first. It is important to appreciate that identifying pertinent levels of development and/or visitation alone is not sufficient to ensure anything. Capacity has long been accepted as being best thought of as a management issue. Of equal importance to identifying critical capacity levels is the management of development to ensure that these levels are not exceeded (Walter 1982; Martin and Uysal 1990; Craik 1995).

A key problem in most tourist destinations is the fact that there is often no management or control of tourism development. Pigram (1990) has noted that much planning of tourism is really marketing, and the responsibility for the management of tourism is more often absent than present in most destinations. The major reason for this is the fact that most tourist destinations are either urban centres (particularly in the case of mass tourism destinations) and well established resorts, or are privately owned and operated establishments, as in the case of integrated large-scale developments.

Only where destinations are part of, or within, areas such as national parks or on public land is there likely to be active management of the development and operation of tourism by individuals or organizations with little or no conflict of interest. While it can be argued that the owners of tourism enterprises have it in their own best interests to manage development and operations with a view to not exceeding capacity and therefore prejudicing the attractiveness of their facility, in fact it would appear that many tourism operators manage their enterprises along the same principles as the pastoralists symbolized in Hardin's (1968) 'The tragedy of the commons'. The possibility of a short-term gain in return for expansion appears to outweigh any concerns over a possible long-term and perhaps permanent decline in resource quality and thus facility viability caused by overuse. This theme is discussed in more detail by the author (1991) with respect to sustainability, and by Healy (1994) in the context of tourism common pool resources.

The 'tragedy' in the context of Hardin's commons was the inevitability of degradation and destruction of the resource, as the only way for users to increase their individual profit was to overuse the resource. In the absence of any controlling or regulatory body, this tended to happen in most situations. Tourism development would appear to be little different in many cases, which is why the inevitability discussed by Christaller (1963) and Plog (1974) and implicit in Butler's article (1980) is significant. Clearly, to achieve sustainability, control and perhaps regulation is likely to be necessary, except in those rare instances where there is complete community and industry agreement to limit development to a level which is sustainable, or within the capacity of the destination. As little satisfactory research has been done on the capacity of tourist destinations (Butler 1996) it is unlikely that such a situation is likely to occur frequently.

Pressures for development

As with most economic activities, there are very real economies of scale in tourism development. Larger facilities are normally more economical per unit to operate than small ones, and thus as business increases most operators wish to expand profits. One way both to increase numbers and reduce relative costs per guest is to increase facility size. In certain situations this may not be a problem. For example, Las Vegas has several of the largest hotels in the world and operators are constantly expanding and redeveloping their facilities. Large size, in a situation such as Las Vegas, does not seem to be a problem in terms of attractivity to potential visitors. Indeed, there appears to be a positive relationship between average hotel size and total numbers of visitors to the city. In other contexts, however, large is not automatically better, and increases in the scale of development may be undesired by traditional visitors. If the destination is to continue to attract visitors and to replace those lost because of displeasure with large-scale facilities, then the destination has to either change its market or to find some way of retaining its original guests. Lower costs and more specific advertising and marketing are the most popular approaches.

Additional or new advertising (in some cases, in the early stages of development there may have been no advertising at all) requires expenditure, and to recoup this, additional tourists are needed. To accommodate extra visitors, additional or larger accommodation units are needed, and in many cases, improved infrastructure, particularly of basic services and transportation. Destinations rapidly find themselves moving in one direction only, that of additional development. To gain additional and improved transportation services and access, particularly from external agencies such as international airline companies, they may need to guarantee a larger and more definite market. The transformation from small-scale, often locally owned development to large-scale, often internationally owned development, is a common phenomenon in many tourism destinations (Pearce 1989).

It should be emphasized here that there is nothing inherently wrong or inappropriate about such a process, however much some proponents of a 'small is beautiful' philosophy may argue. It is clear, however, that where such a transformation may occur, if it is not intentional, not desired by locals and not well thought out, then ultimately it is likely to bring about dissatisfaction and a range of additional problems. Even where the above conditions are met, it is likely that the changes brought about by large-scale tourism development will create some unanticipated impacts and difficulties, as well as significant changes to the destination. Of equal concern is the probable effect of such development upon the original tourists to a destination. As with consumers of many services, the deep and basic motivations that push or pull tourists to a destination are not well understood (Pearce 1993), but destinations which have moved from small-scale development to large-scale development may not retain many of their original customers. Whatever the limitations of Plog's (1974) model, there is an intuitive truth about his segmentation of tourists. His 'allocentrics' (jet-setters), and Christaller's (1963) *jeunesse dorée* leave destinations rapidly and finally once they pass a certain level of development, if that development results in a significantly enlarged market. This may well be a form of elitism, simplistically a manifestation of 'class preceding mass', or may be a more deeply felt clash of tastes and preferences. In many wilderness areas in North America, where access has been improved and numbers of visitors increased, the satisfaction of those desiring what they perceive as a true wilderness experience, incorporating solitude and the relative absence of other visitors, has long been shown to decline in proportion to an increase in an undesired type and quantity of visitors (Lucas 1964). It would be surprising if a similar process did not take place at tourist destinations, and in fact it is clear that it does. As Wheeler (1993) has so pointedly illustrated, exclusivity takes many forms, and sensitive ecotourists are no less selective and elitist in their preferences than other types of tourists. The transformation through development of destinations in areas such as the Scottish Highlands have been well documented, and with those changes have come significant changes in the nature of the visitors (Butler 1985; Getz 1992).

Thus it is hard to avoid the conclusion that development, however sensitive, ultimately will change the mix of tourists coming to a destination. One may sketch a simplistic scenario. The mix of tourists is unlikely to be a contented one, since at the extremes they represent different preferences and tastes. The 'old' tourist will be longing for things to remain 'as they were', or more as they were perceived when they first came to the location (a little like the 'era' approach to national park planning), whereas 'new' tourists are likely to want some additional development if they are to return. Faced with potentially losing both types of visitor, most destinations and their developers will opt for additional development, anticipating that this will not only ensure the return of the 'new' tourists, but will also attract more of them, thus increasing return on investment. Such a development will make the destination increasingly different and unattractive to any remaining 'old' tourists. In line with Plog's (1974) model, the destination will inevitably move through the market from allocentric to pyschocentric. Initially there will be an increased market, but ultimately the market will peak and decline, as the destination runs through its life cycle.

This can be viewed as a deterministic and unidirectional view, which makes no allowance for the dynamics of the tourist industry, nor for actions of local communities, of governments, and of external forces. There is certainly validity in these criticisms, but the fact remains that in reality, few destinations do not follow such a pattern of development. Almost all of the pressures tend to make it attractive and desirable to do so. The trick is clearly to achieve one of two things: unlimited and indefinite growth, or sustainability. The latter is difficult, and perhaps unattainable; the former is impossible.

EVOLUTION AND SUSTAINABILITY

Perhaps the key problem with sustainable development in tourism is that the concept has been adopted with so much enthusiasm but so little real action. Entrepreneurs are claiming to operate sustainable tourism facilities even before they are open for business. They beget the fundamental question, namely: what is sustainable development in the context of tourism? This author has discussed this issue elsewhere (Butler 1993) and argued for a distinction between sustainable tourism and sustainable development of tourism. This is not a matter of mere semantics. For almost two centuries Niagara Falls has had a reasonably successful tourism industry, which could be justly called sustainable. However, no-one is likely to hold up Niagara Falls as an example of sustainable development. Since its inception Las Vegas has annually attracted ever more visitors, despite being so blatantly unsustainable in an environmental sense. Such places are sustainable in the sense of continually being able to attract tourists because they make little attempt to rely on natural and inherent features and have made the leap to attractions based almost entirely on created elements. The fantasy worlds of Las Vegas, like those of Disneyworld and Disneyland, are attractive because they are so different from the natural world that escape from the mundane is effortless, and they retain that attractivity by continually altering and increasing the range of attractions. Ultimately, one assumes, they will run out of attractions, energy, space or money, or some combination of the above, and visitation will decline. Niagara Falls appears to be close to that situation at present, but a casino opened at the end of 1996, and rejuvenation, for a time at least (as with Atlantic City) may take place (De Meel 1996). The original attraction, the Falls, now no longer a natural feature, consumes perhaps at a maximum 10 per cent of the average day visitor's time in Niagara Falls, and a minuscule portion of the visit of overnight guests. Las Vegas, perhaps more than anywhere, symbolizes Wolfe's 'divorce from the geographic environment', although in the case of Las Vegas it could be argued that the geographic environment was never an attraction in the first place.

It is tempting to argue further that as a tourist destination evolves, it must inevitably move further from any semblance of sustainability, and given the discussion above, that no destination could remain sustainable, if it ever had been in such a state. Such an argument is probably true, but there seems to be no really acceptable definition of sustainable development in the context of tourism, and few places on earth are truly sustainable. If sustainability does exist, it can probably only do so at the global scale, and even here the process is under stress. Thus one may argue that it is unrealistic to expect destinations to be sustainable in themselves; indeed, Wall (1996) has questioned whether even ecotourism, supposedly the most benign form of tourism, can be sustainable. Rather, a more important issue is whether tourism destinations can contribute to sustainable development ideals. The original use of the term sustainable development (WCED 1987: 43) involved meeting 'the needs of the present without compromising the ability of future generations to meet their own needs'. In the context of tourism there is no clear understanding of what the 'needs' of the present generation are, let alone comprehending what the needs of future generations may be. Thus perhaps the best that can be hoped for is for tourist destinations to fill niches in parts of the overall collection of needs of this generation and the needs of future generations by remaining attractive to tourists over the long term. It is unrealistic to expect destinations and developments to be individually sustainable in an environmental sense.

One of the key difficulties facing most tourist destinations, particularly the older destinations, is that they are not easily capable of becoming sustainable because of the inherently dynamic nature of tourism and the tourist market discussed above. Most destinations, other than those created from nothing in the last decade or so, have evolved, and contain remnants of previous lives, either as communities not involved with tourism, or involved with a different form of tourism, and are in a constant state of change. Many of the older ones have still not yet completely accepted the automobile, and find many of their attractions are suffering because they were designed for or modified to accommodate a different type of tourist practising a different type of tourism. The major problem facing tourism development is not ensuring new development is environmentally sustainable, but in attempting to make existing developments and destinations sustainable well into the future, and thus capable of meeting the needs of future generations. They are unlikely to achieve this by following unchecked evolution for the reasons noted earlier. Once a destination has been developed so greatly or so rapidly that it has exceeded its innate capacities, in terms of factors such as its ability to accommodate economic expansion and visitor expenditures, physical demands for space, demands for other natural resources such as water, beaches, and wildlife, and the tolerance of local residents to change and disturbance, it is unlikely to be able to remain attractive to generations of future tourists without massive investment and redevelopment. The attitude and actions of the tourism industry and many elements of the public sector in many countries have meant that it is easier, cheaper and more beneficial in the short term to develop a new destination than it is to redevelop and renovate an old one. In the future, if not the present, such attitudes will have to be found as unacceptable as they are unsustainable.

CONCLUSIONS

The process of development of tourist destinations has received relatively little attention in the literature, and conceptualization of the process has been very limited. There have been a large number of case studies of the pattern of development of destinations, but they have been based on a shallow theoretical foundation. It has been argued here that most tourist destinations experience an evolutionary process, although it is recognized that in the last decade or so there have been destinations established on a more revolutionary pattern, represented by sudden, rapid and major development quite different from what had existed in that location previously. The majority of destinations, however, have evolved from settlements and functions not related to tourism, but which have possessed features and attractions which have proved capable of drawing visitors. A model of the life cycle of destinations (Butler 1980) which incorporates the concept of carrying capacity is still felt to be a generally accurate, if in some cases incomplete, representation of reality for most destinations.

Tourism is extremely dynamic, as are all the elements which comprise it, and few destinations are able or wish to remain unchanged. By adapting to accommodate changing preferences and types of visitors, destinations hope to remain viable and sustainable in tourism into the future. It is essential, however, that such change does not include such an overtaxing and subsequent degradation of the resources which first attracted tourists that the destination loses all semblance of its former self, for few such destinations have managed to survive such a process in the extremely competitive industry that is tourism. Destinations which rely totally on completely contrived attractions are more vulnerable than most to competition, since such attractions can most often be developed at any location. Given the attraction to most consumers of new items and facilities over existing ones, new developments are likely to attract the market away from existing destinations. The only hope for such destinations to avoid almost inevitable decline is to continue to provide ever more fantastic and contrived attractions, subscribing to the 'dreadnought' philosophy practised by major casinos in the United States (Stansfield 1996) and visible in Las Vegas, and to a lesser degree Atlantic City. Ultimately, the limit will be reached, and it is then likely that uncompetitiveness and decline will result. A destination may then have the difficult choice of leaving tourism entirely, requiring a massive change in image projection and possibly infrastructure provision, as well as labour force retraining, or an equally massive change in marketing direction, in an attempt to attract a new form of tourism or leisure. It is clearly in a destination's best interest to try to remain attractive to tourists, and therefore sustainable, as long as possible. Anticipatory planning, community support, clarification of the effects and changes associated with tourism compared with other possible forms of development, and regulation and control of the rate, type and level of development, are all essential if destinations are to move towards sustainable development. Even then, the dynamics of tourism are likely to ensure that sustainability can only be approached, rather than permanently achieved.

REFERENCES

Agarwal, S. (1994) 'The resort cycle revisited: implications for resorts', in C. P. Cooper and A. Lockwood (eds) *Progress in Tourism, Recreation and Hospitality Management*, vol. 5, Chichester: Wiley.

Bryden, J. M. (1973) *Tourism and Development: A Case Study of the Caribbean*, Cambridge: Cambridge University Press.

Butler, R. W. (1980) 'The concept of a tourist area cycle of evolution: implications for management of resources', *The Canadian Geographer*, 24 (1): 5–12.

——(1985) 'Evolution of tourism in the Scottish Highlands', *Annals of Tourism Research*, 12 (3): 371–92.

——(1989) 'Tourism and tourism research', in T. L. Burton and E. L. Jackson (eds) *Understanding Recreation and Leisure: Mapping the Past, Charting the Future*, State College PA: Venture Publishing, 567–95.

——(1991) 'Tourism, environment and sustainable development', *Environmental Conservation*, 18 (3): 201–9.

——(1993) 'Tourism: an evolutionary perspective', in J. G. Nelson, R. W. Butler and G. Wall (eds) *Tourism and Sustainable Development: Monitoring, Planning, Managing,* Waterloo, Canada: University of Waterloo, 27–43.

——(1996) 'The concept of carrying capacity for tourist destinations', *Progress in Tourism and Hospitality Research,* 2, 3/4, 283–92.

Choy, D. J. L. (1992) 'Life cycle models for Pacific island destinations', *Journal of Travel Research,* 30 (3): 26–31.

Christaller, W. (1963) 'Some considerations of tourism location in Europe: the peripheral regions; underdeveloped countries; recreation areas', *Regional Science Association Papers,* 12: 103–18.

Clawson, M. (1959) *The Crisis in Outdoor Recreation,* Washington DC: Resources for the Future.

Cooper, C. (1994) 'The destination life cycle: an update', in A. N. Seaton (ed.) *Tourism: State of the Art,* Chichester: Wiley: 340–6.

Cooper, C. and Jackson, S. (1989) 'Destination life cycle: the Isle of Man case study', *Annals of Tourism Research,* 16 (3): 377–98.

Craik, J. (1995) 'Are there cultural limits to tourism?', *Journal of Sustainable Tourism,* 3 (2): 87–98.

Debbage, K. G. (1990) 'Oligopoly and the resort cycle in the Bahamas', *Annals of Tourism Research,* 17, 513–27.

De Kadt, E. (1973) *Tourism: Passport to Development?,* New York: Oxford University Press.

De Meel, S. (1996) 'Resident attitude and behaviour toward tourism in New York', unpublished masters thesis, University of Western Ontario.

Di Benedetto, C. A. and Bojanic, D. C. (1993) 'Tourism area life cycle extensions', *Annals of Tourism Research,* 20: 557–70.

Din, K. H. (1992) 'The "involvement stage" in the evolution of a tourist destination', *Tourism Recreation Research,* 17 (1): 10–20.

Getz, D. (1983) 'Capacity to absorb tourism: concepts and implications for strategic planning', *Annals of Tourism Research,* 10: 239–63.

——(1992) 'Tourism planning and the destination life cycle', *Annals of Tourism Research,* 19 (4): 752–70.

Gilbert, E. W. (1939) 'The growth of island and seaside health resorts in England', *Scottish Geographical Magazine,* 55: 16–35.

Hardin, G. (1968) 'The tragedy of the commons', *Science,* 162: 1243–8.

Hartman, R. (1986) 'Tourism: seasonality and social change', *Leisure Studies,* 5 (1): 25–33.

Haywood, K. M. (1986) 'Can the tourist-area life cycle be made operational?', *Tourism Management,* 7: 154–67.

——(1992) 'Revisiting the resort cycle', *Annals of Tourism Research,* 19 (2): 351–4.

Healy, R. G. (1994) 'The "Common Pool" problem in tourism landscapes', *Annals of Tourism Research,* 21 (3): 596–611.

Hovinen, G. R. (1982) 'Visitor cycles: outlook in tourism in Lancaster County', *Annals of Tourism Research,* 9 (3): 565–83.

Ioannides, D. (1992) 'Tourism development agents: the Cypriot resort cycle', *Annals of Tourism Research,* 19 (4): 711–21.

Johnson, P. and Thomas, B. (1994) 'The notion of capacity in tourism: a review of the issues', in C. P. Cooper and A. Lockwood (eds) *Progress in Tourism, Recreation and Hospitality Management,* vol. 5, Chichester: Wiley.

Kermath, B. M. and Thomas, R. N. (1992) 'Spatial dynamics of resorts: Sosua, Dominican Republic', *Annals of Tourism Research,* 19: 173–90.

Lucas, R. C. (1964) 'Wilderness perception and use: the example of the Boundary Waters Canoe Area', *Natural Resources Journal*, (3): 394–411.

Lundgren, J. (1973) 'The development of the tourist travel systems', *Tourist Review*, 1: 2–14.

——(1983) 'Development patterns and lessons in the Montreal Laurentians', in P. Murphy (ed.) *Tourism Canada: Selected Issues and Options*, Western Geographical Series, Victoria: University of Victoria, 95–126.

Martin, B. S. and Uysal, M. (1990) 'An examination of the relationship between carrying capacity and the tourism life cycle: management and policy implications', *Journal of Environmental Management*, 31: 327–33.

Mathieson, A. and Wall, G. (1982) *Tourism: Economic, Physical and Social Impacts*, New York: Longman.

Meadows, D. H., Meadows, D. L, Randers, J. and Behrens, W. W. (1972) *Limits to Growth*, New York: Universal Books.

Meyer–Arendt, K. J. (1985) 'The Grand Isle, Louisiana resort cycle', *Annals of Tourism Research*, 12 (3): 449–65.

Miossec, J. M. (1977) 'Un modèle de l'éspace touristique', *L'Espace Géographique*, 6 (1): 41–8.

Nelson, J. G., Butler, R. W. and Wall, G. (1993) *Tourism and Sustainable Development: Monitoring, Planning, Managing*, Waterloo, Canada: University of Waterloo.

Noronho, R. (1976) *Review of Sociological Literature on Tourism*, New York: World Bank.

Ogilvie, F. W. (1933) *The Tourist Movement*, London: Staples Press.

Oglethorpe, M. (1984) 'Tourism in Malta', *Leisure Studies*, 3: 147–62.

Pearce, D. G. (1989) *Tourist Development*, Harlow: Longman.

Pearce, P. L. (1993) 'Fundamentals of tourist motivation', in D. G. Pearce and R. W. Butler (eds) *Tourism Research: Critiques and Challenges*, London: Routledge, 113–34.

Pigram, J. J. (1990) 'Sustainable tourism: policy considerations', *Journal of Tourism Studies*, 1 (2): 2–9.

Plog, S. C. (1974) 'Why destination areas rise and fall in popularity', *Cornell Hotel and Restaurant Administration Quarterly*, 14: 55–8.

Prosser, G. (1995) 'Tourist destination life cycles: progress, problems and prospects', paper presented at National Tourism Research Conference, February 1995, Melbourne, Australia.

Smith, R. A. (1992) 'Beach resort evolution: implications for planning', *Annals of Tourism Research*, 19: 304–22.

Smith, V. (ed.) (1977) *Hosts and Guests: The Anthropology of Tourism*, Philadelphia: University of Pennsylvania Press.

Stansfield, C. (1978) 'Atlantic City and the resort cycle', *Annals of Tourism Research*, 5 (2): 238–51.

——(1996) 'Reservations and gambling: Native Americans and the diffusion of legalized gambling', in R. W. Butler and T. Hinch (eds) *Tourism and Indigenous Peoples*, London: Routledge, 129–49.

Strapp, J. D. (1988) 'The resort cycle and second homes', *Annals of Tourism Research*, 15 (4): 504–16.

Towner, J. (1996) *An Historical Geography of Recreation and Tourism, 1540–1940*, Chichester: Wiley.

Wall, G. (1996) 'Is ecotourism sustainable?', *Environmental Management*, 2, 3/4, 207–16.

Walter, J. A. (1982) 'Social limits to tourism', *Leisure Studies*, 1, 2: 295–304.

Weaver, D. (1988) 'The evolution of a "plantation" tourism landscape on the Caribbean island of Antigua', *Tijdschrift voor Economische en Sociale Geografie*, 79: 319–31.

——(1990) 'Grand Cayman Island and the resort cycle concept', *Journal of Travel Research*, 29 (2): 9–15.

Wheeller, B. (1993) 'Sustaining the ego', *Journal of Sustainable Tourism*, 1 (2): 121–9.

Williams, M. T. (1993) 'An expansion of the tourist site cycle model: the case of Minorca (Spain)', *Journal of Tourism Studies*, 4 (2): 24–32.

Wolfe, R. I. (1952) 'Wasaga Beach: the divorce from the geographic environment', *The Canadian Geographer*, 2: 57–66.

——(1966) 'Recreational travel: the new migration', *The Canadian Geographer*, X (1): 1–14.

World Commission on Environment and Development (1987) *Our Common Future*, Oxford: Oxford University Press.

Young, G. (1973) *Tourism: Blessing or Blight?*, Harmondsworth: Penguin.

Part III

BALANCING TOURISM GROWTH WITH SUSTAINABILITY

8

SUSTAINABLE TOURISM IN THE DEVELOPING WORLD

Salah Wahab

INTRODUCTION

Many developing countries (or less developed countries or Third World countries) seem to suffer generally from external indebtedness, scarcity of foreign currency earnings, under-utilization of some of their major resources, comparatively disadvantageous exports, inadequate development finance and poor quality of life. Falling prices for the commodities dominating their economies exert a profound negative impact upon production and employment ratios in such countries. One comparative advantage that developing countries seem to enjoy, is their still unspoiled nature and their attractive and genuine, though not necessarily modern, way of life.

Policies affecting trade, the flow of capital, revenues and global finances, as well as environmental concerns, have a major impact upon sustainable development. Agenda 21 adopted by the United Nations Conference on Environment and Development on 14 June 1992 is the international community's response to the need for devising strategies to halt and reverse the unwarranted effects of poverty and environmental degradation. Liberalization of trade, making trade and environment mutually supportive, providing adequate financial resources, dealing equitably with international debt, and encouraging macroeconomic policies conducive to environment and development, are some of the essential solutions proposed in Agenda 21 to upgrade the conditions prevailing in developing countries.

TOURISM AND DEVELOPMENT

In 1973, Robert Erbes commented that everything seems to suggest that developing countries look upon tourism consumption as manna from heaven that can provide a solution to all their foreign settlement difficulties. Although this statement was considered by some authors as a simplistic presumption (Jenkins 1992) it is, however, a useful introduction to the topic of tourism and development.

Tourism has noticeably progressed to become a major economic activity

(De Kadt 1979: 3) and a world economic activity that has proved its vital and irreplaceable role in world trade (Wahab 1975: 52). However, tourism in the context of the national economy 'is not a unique devil' as De Kadt stresses (1979: 12).

> The ability of the national economy to benefit from tourism depends on the availability of investment to develop the necessary infrastructure and on its ability to supply the needs of tourists ... etc. In general, small economies are more likely to be dependent on imports, and the same is true of economies where tourism is not very well developed.
>
> (Williams and Shaw 1988: 5)

As suggested by some writers, 'there is no other international activity which involves such critical interplay among economic, political, environmental, and social elements as tourism' (Lea 1988: 2). However, Young (1973) points out that the degree of acceptable reliance on tourism relates to the structure of the economy:

> Where there is high unemployment, a relatively unskilled labour force and few alternative sources of employment ... then stimulation of the tourist industry may well be a correct course of action. The danger appears to be at the next stage of economic development, where unemployment and under-employment have been reduced, the labour force is better educated and an infrastructure exists which might support other industries. Continuing dependence and emphasis on tourism may no longer be economically justifiable.
>
> (Young 1973: 61)

Tourism and economic development

It is generally recognized that tourism stimulates the development of several sectors of the national economy. Tourism:

- creates new local requirements for equipment, food and other supplies fostering new industries and commercial activities and creating a new market for them,
- has a favourable impact upon employment in a country as it increases the opportunities available for work in accommodation, food industries, tour operations and travel agencies, government tourist offices, handicraft and souvenir trades, recreational, amusement, and entertainment activities, and various selling outlets,
- increases urbanization through the continuous growth of construction and renovation of tourist facilities. This implies creating and improving infrastructures and tourist superstructures, particularly in remote and depressed areas,

- helps to increase the state earnings of hard currency necessary for bridging or reducing whatever deficit there is in the balance of payments, and thus fosters the development of the national economy,
- is one of the most effective redistributive factors in international economic relations. To explain this, one has to admit that travel is a social activity arising from surplus income. Thus the flow of foreign travel is more significant from richer countries to the less privileged. A substantial portion of foreign travel is directed towards developing regions which are more attractive naturally as they are not yet spoiled by industrialization. Thus tourism redistributes capital between developed and developing countries, and
- activates the economic circuit in a country, thus accelerating the multiplier effect.

(Wahab 1974: 15)

In contrast to the undeniable economic benefits that can accrue to the tourist destination, tourism also exacts a price from it. Such a price is represented by investments in tourism infrastructures and superstructures, in the cost in foreign exchange of required imports, and other aspects such as distortion of the national value system, vandalism, congestion, increased crime, and the possible change or loss of the national architectural heritage and natural beauty (Wahab 1993).

In developing countries with good tourism potential, tourism may be adopted as an economic activity that could rationalize economic policy, especially through balanced growth brought about by new or additional business production cycles prompted by tourism expansion. Thus tourism can become the cause and the effect of rationalizing economic development, along with the economic production sectors of agriculture and industry (Wahab 1993).

Tourism and social development

Insofar as social development is concerned, tourism as a phenomenon represents an important leisure activity which has to be included in the state's national policy, in the interests of safeguarding its moral and health standards. Moreover, tourism is a factor of social solidarity and equilibrium. The movement of tourists, whether domestic or international, across the different regions of the destination, makes it easier to bring about social homogeneity among the different ethnic, religious, urban and rural groupings. In addition, human contacts arising from tourism yield more social benefits than any other industry, in particular, developing new tastes and modern habits, and fostering cultural borrowings (Wahab 1974: 14).

However, strange habits, values, heterogeneous traditions and different mentalities may bring about a cultural shock which might lead to social problems (Wahab 1974: 14).

Tourism's positive contribution to the economic, social, political and environmental advancement of developing countries is contingent upon the able implementation of suitable scientific, technical and technological factors, as well as the appropriate deployment of available resources to maximize benefits and minimize disadvantages. This is not an easy task given that developing countries usually lack adequate expertise. Such weakness could be surmounted by bilateral assistance from developed countries, as well as international cooperation through international organizations, namely WTO, ILO, the World Bank and Regional Development Banks.

Both proponents and opponents of tourism as a tool of development in developing countries admit that tourism has made influential changes in the structure, value systems, traditions and behavioural patterns of many societies.

The debate about whether these changes are good or bad is still unresolved, as the interests of society and the individual are not necessarily identical or even similar (Jenkins 1992). Moreover, the varying conditions prevailing in different societies make generalizations difficult, if not impossible, to reach.

Tourism and the environment

As rightly put by Pigram, 'tourism is, to a large degree, a resource-based activity, interacting with natural systems and with a capacity to initiate far-reaching changes on the environment' (Pigram 1995: 208).

Therefore, management of environmental issues has become central to tourism planning in all destinations, developed and developing, and thus is a basic premise for sustainability. There is a long catalogue of environmental damage caused by various human activities, although tourism development projects are not the only offenders (Jenkins 1992). Avoidance of such environmental damage in tourism must be given priority. Protected natural areas and national parks should be encouraged in environmentally sensitive areas. In these areas, development should be prohibited or at least restrained. Unique cultural attractions such as the Giza Pyramids and the Sphinx in Egypt, Barubudur Temple in Central Java and Matsu Pichu in Peru should be properly protected and maintained as they are priceless examples of human heritage. A basic limitation is the carrying capacity of these areas and attractions in relation to visitor use and the development of facilities. Tourism should be used as a medium to attain such conservation in order to maintain resource sustainability. Otherwise the environmental damage will adversely affect resource quality and tourists will avoid the destination.

Any alleged disregard for the environment by conventional tourism development typically yields expressions of outrage. Hong expresses this mood:

Having ruined their own environment, having either used up or destroyed all that is natural, people from the advanced consumer societies are compelled to look for natural wildlife, cleaner air, lush greenery and golden beaches elsewhere. In other words, they look for other environments to consume. Thus armed with their bags, tourists proceed to consume the environment in the countries of the third world – that last 'unspoiled corner of earth'.

(Hong 1985: 12)

TOURISM DEVELOPMENT POLICY

Coherent policy conception, formulation and implementation are not yet well structured in most developing countries. This is particularly true in tourism which is a multifaceted industry requiring a good deal of coordination, organization, planning, motivation, sound utilization of resources and proper implementation. Whereas legitimate demands on government are usually large in developing countries, the capacity for sound public management of available and potential resources is not usually well developed. This results in an *ad hoc* reactive approach to solving problems that impose themselves on the community, instead of a systematic policy that typifies proactive approaches.

General tourism policies are those policies which serve as a framework specifying the national tourism goals, objectives, priorities and actions that will provide the basis for future development of tourism in the destination area. In other words, tourism policy determines the climate in which the country functions economically, socially, culturally, politically and environmentally (Theuns 1987: 14). As Sessa observed, tourism policy, 'as an integral part of a nation's overall economic policy, must be coordinated with the policies of all other sectors directly or indirectly related to tourism' (Sessa 1976: 26).

Despite the indispensability of tourism policy as a precursor for future tourism planning and development, many developing countries have yet to develop statements of their tourism policies. Typical tourism goals are:

- *economic goals*, meaning optimization of the contribution of tourism and recreation to economic prosperity: full employment, regional economic development and improved international balance of payments;
- *sociocultural goals*, represented by (1) the personal growth and education of the population and the boosting of their appreciation of the history, geography, and ethnic diversity of the country; (2) avoidance of any activities that may undermine or denigrate the social and cultural values and resources in the country or area as well as negatively affecting its traditions and/or lifestyles; (3) maximizing the chances for a more beneficial enjoyment of travel and recreation for foreign visitors and residents;

133

- *environmental goals*, oriented towards (1) judicious use of natural resources; (2) avoidance of all possible causes of pollution; (3) utilization of new and renewable sources of energy; (4) safeguarding the physical environment through strict adherence to carrying capacities and clearance of solid waste; and (5) preservation of national heritage resources and urban revitalization. Environmental protection is one means to an end which is sustainability;
- *market development goals*, represented by (1) frontier facilitation procedures; (2) increasing the chances of a better tourist image of the destination in generating markets; and (3) enhancing the opportunities for a broader market for the national tourist products; and
- *government operations goals*, which include (1) maximum harmonization of all government activities supporting or relating to tourism and recreation; (2) supporting the need to educate all policy makers on tourism; (3) legislation for necessary tourism activities; (4) regulation of the various facets of tourist action at the national, regional and local levels; (5) the raising of tourism-consciousness amongst the general public; (6) encouraging the private sector to increase its activities in tourism development through various incentives; (7) ascertaining the limits of tourism growth; and (8) safeguarding internal and external security in tourism operations.

Tourism strategies and programmes

Having established the tourism policy (in one form or other, e.g. as a set of basic tourism laws), various tourism strategies and programmes are needed to achieve the policy's goals and objectives. Examples of such strategies include specifics of the tourism facilitation strategies, investment incentives, development research, marketing research, priority tourism development areas and zones, marketing and promotional strategies in various niche markets including domestic tourism, air transport and cruise strategies, and a tourism education and training strategy.

The question of quality versus quantity appears at the forefront of tourism development strategies. Besides the prominence given to quality as a major tourist destination attraction at the macro- and micro-levels, the logical inclination may well be to wish to attract the visitor with the highest daily expenditure who accordingly adds most to the national income (unless the additional expenditure is outweighed by a greater volume of imports), the visitor with special interest in culture, or the incentive visitor.

Carrying capacity is a central principle in any strategy for sustainable tourism development. It determines the maximum use of any place without causing negative effects on the resources, on the community, economy, culture and environment, or reducing visitor satisfaction. It is virtually a limitation on ill-conceived tourism growth.

The respective roles of the public and private sectors as entrepreneurs in tourism are matters of a political decision in the light of a country's economic and social system, forming an integral part of the tourism development policy (UNCTAD 1973).

Community participation

Community participation in the decision-making process of tourism development is a key issue in ensuring the acceptability of tourism, thus contributing to its sustainability. This derives not only from fair and just rules of democracy, but also from the fact that tourism should not expand at a rate beyond which citizens in a given community actually desire and can control. Acceptance by consensus would not be required. However, suffice to say that in order to channel the social impact of tourism in its proper path, tourism should not cause disillusionment to the host society. A model that was based on research in Barbados developed an irritation index of five stages, starting with initial euphoria, then apathy, increasing irritation, outright antagonism, and finally a stage when cherished values are forgotten and the environment destroyed by mass tourism (Lea 1988: 64).

Community participation in the tourism development process differs in developing countries from developed countries. By and large, such difference manifests itself in three ways. *First,* local communities in some developing countries devote minor attention to issues of tourism development and planning as they are much more troubled by the lack of clean and hygienic food and drink and suitable shelter in the short time frame. *Second,* lack of democracy in many developing nations dictates that the will of the ruling class expresses the *pro bono publico* (the public good). *Third,* there is no system that would allow social outputs to be determined by the people most immediately affected by them, even in the presence of local government, as tourism is usually looked upon as an industry of national concern.

Community attitudes towards tourism might change in time, e.g. from enthusiasm to euphoria or apathy when tourism fails to live up to its promise over a reasonable period of time. This may also happen when the community leadership changes over time.

A typical example of such change in community attitudes is the case of Luxor's visitor centre in Egypt. In 1983, a 'Study on visitor management and associated investments on the West Bank of the Nile at Luxor' recommended *inter alia* the creation of a visitor centre to meet visitors' needs for tourist information and for services and amenities. The primary function of the visitor centre was to serve as the central focus of an information interpretation programme. Another equally important function was to house priority tourist amenities – specifically a restaurant, theatre and shops. The Luxor West Bank of the Nile encompasses the Valley of Kings and the Valley of the Queens and Nobles, in addition to some important temples (Der al-Bahari) and monuments. Other organizational and management recommendations were intended to reduce pressure on the most-visited tombs, which were in danger of deterioration.

Consultations with the Government Archaeology Department (at that time) and the Luxor Community represented by the City Council resulted in unanimous endorsement of the project in 1986.

When the visitor centre was constructed seven years later, an opposition front soon appeared: the Luxor Business Community became antagonistic to the visitor centre and the whole of the proposed Luxor West Bank organization and management system came to a halt. This was due to the change in leadership of the Luxor Business Community, which considered that the plan was against the local population's recognized interests.

TOURISM DEVELOPMENT STRATEGIES AND SUSTAINABILITY

Principles and case studies

Tourism, as a vehicle for development in developing countries, should be an *evolutionary* process of change to a better future. It should create a *structure* that aims to create a steady and balanced *rate of change*. Such rate of change has to be consistent with the prevailing socioeconomic, politico-cultural, educational, organizational and environmental conditions in the destination, because development starts with people and not with goods. As rightly raised by Schumacher, every country, no matter how devastated, which had a high level of education, organization and discipline, produced an 'economic miracle' (Schumacher 1973: 157).

Demands for a growing volume of tourist traffic can affect a country's resources sustainability potential. Nevertheless, sustainability is still a new concept and a costly notion to implement in developing countries. These countries cannot normally shoulder the costs of environmental management (Erlet Cater 1993) and therefore they have to approach it in a gradual manner. The only blessing they enjoy is that their physical environments are still comparatively natural, and therefore the pertinent cost should not be excessive.

While this is true in many cases, some writers consider that as most tourists arriving in developing countries are unfamiliar with local conditions and unaware of finely balanced ecological systems, they are potentially, if unintentionally, dislocative and destructive. Even the ecotourists' motto, 'Leave nothing but your footsteps, take nothing but your memories', needs to be amended in the face of, for example, the rainless desert of southern Namibia. There, 'footsteps in the sand, and 4-wheel drive vehicle tracks, can scar the landscape for years' (Doggart and Doggart 1996: 71–86).

This is an exaggerated view on the part of conservation biologists, which if taken into consideration in sustainable tourism development, would mean that no tourism activity would be warranted in developing countries.

Competitive tourism strategies are becoming critically important for developing countries because tourism, by its nature, is a sensitive and highly competitive industry. While natural attributes form distinctive advantages for developing nations, and there is a growing trend towards nature-based tourism, the use of human intelligence and creativity, and the application of scientific knowledge and technology can make the difference. Cost-effectiveness in tourism investments is a must and not a luxury in developing countries. Total quality management has become another limitation that conditions sound development planning from which sustainability can ensue. The quality of services delivered by the tourism inner circle staff (e.g. hotel staff, travel agency staff, tour guides, taxi and tour-bus drivers, river-cruise staff) is of prime importance. Equally important is the protection and conservation of natural and cultural assets, the cleanliness of the whole destination, necessary hygiene in food and drinks, friendliness of the people, innovation in product development, creativity in marketing strategies, access to the proper distribution channels in the market place, suitable use of technologically advanced information systems, and the capability to anticipate and accommodate change. All this could make tourism a lead sector in a country's developmental strategy (Poon 1993).

But tourism cannot and should not be developed in isolation from other productive and service industries in the destination. It is an industry that interacts with many sectors and a multitude of government functions, and should therefore be developed with a view to such multidimensional characteristics. Adoption and implementation of new development and marketing strategies, activation of the main sector-related issues (e.g. research, market surveying and sourcing of materials), development of an environmental focus, encouragement of the private sector through various incentives to become a dynamic sector, guaranteeing adequate accessibility, human resources development, and transformation of the role of national tourist administrations in the market place to become active and effective image builders, are some of the main tools with which to build a sustainable tourism development programme.

Needless to say, a short-sighted approach in order to achieve maximum economic returns from tourism in the short term may cause detrimental repercussions in developing countries. This approach manifests itself in prices that are disproportionate with service quality, and results in tourists' dissatisfaction and a potential decrease in tourist traffic.

The charter of sustainable development adopted at the World Conference on Sustainable Tourism, which met at Lanzarote (Canary Islands, Spain) on 27–8 April 1995, is significant in providing guidelines for the various governments seeking to ensure the creation of a sustainable tourism industry.

Case study: Egypt

Tourism in the economy

Egypt is one of the oldest civilizations in the world and is mentioned in Biblical and Quoranic texts. Being endowed with historical monuments, archaeological sites and landmarks covering several millennia of civilization, as well as extensive beautiful beaches, it has considerable potential for developing its tourism sector. Besides its diversified topographical and cultural attractions, Egypt's tourist endowments are further enhanced by mild weather for at least eight months of the year.

The Egyptian tourism sector initially received substantial attention. However, this support softened as the development emphasis shifted to the industrial sector. Deterioration of the tourism industry was further aggravated by the war of 1967 and continued hostilities for about ten years until the Peace Agreement. During this period the influx of tourists to Egypt was substantially impaired. A change in economic philosophy, coupled with the signing of the Peace Agreement in the mid-seventies, brought with it increased emphasis on tourism. The number of hotels increased, facilities for servicing the tourism sector improved, and as a result the number of arrivals rose substantially relative to previous decades. This was further assisted by a liberalized civil aviation policy and the diversification of Egyptian tourism from a cultural destination to a combined leisure and cultural attraction.

Between 1985 and 1989, tourism in Egypt, as measured by the number of international arrivals, grew on average by about 13.6 per cent per annum, much higher than world tourism, which grew at an average annual rate of about 5.9 per cent in the same period. From being a modest contributor to foreign exchange earnings of about US$700 million in 1986, tourism receipts reached $1.5 billion in 1990 and $2.8 billion in 1995, putting the industry in second place after workers' remittances, but ahead of Suez Canal revenues and revenues from oil exports. In addition, Egypt has considerably more domestic control over the generation of foreign exchange through tourism, which has shown a greater responsiveness to favourable domestic policies than other sources of foreign exchange. As an illustration, the 37 per cent exchange rate depreciation in May 1987 coincided with a 28 per cent increase in tourist arrivals, a 34 per cent increase in nights spent in Egypt and a 133 per cent increase in recorded tourist receipts in 1987–8 over the previous year.

Apart from being an important foreign exchange earner, tourism development also has a substantial impact on employment in other sectors such as transport, food industry, textiles, the air-conditioning industry, the furniture industry, cottage industries and the informal, small-scale sectors of manufacturing and services. This is especially relevant in view of the current problem of growing unemployment and the need to redeploy redundant labour from government- and publicly owned enterprises. Direct and indirect employment in the tourism sector is estimated at 350,000. This is based on estimates of 110,000 employed in the hotel sector, 25,000 in travel agencies, 10,000 in restaurants, 50,000 in tourism-related retail trade, and some 4,500 tour guides, the remaining 150,000 or so being indirectly employed. About 20 per cent of the workforce in tourism are women.

Although tourism still contributes less than 3 per cent of GDP, it has been since 1987 the fastest growing sector in the economy, in response to favourable economic reforms at both macro- and sectoral levels. The opportunities for increasing its contribution to the economy are excellent, given the country's unique but under-exploited tourist potential.

The tourism market

Tourism to Egypt consists of two separate markets: Arab and foreign. The Arab market comprises visitors from Saudi Arabia, Kuwait, Libya and other Arab countries, who account for about 38 per cent of Egypt's tourism market. The foreign market is dominated by visitors from Germany, France, the United Kingdom and Italy. Only in three long-haul markets has Egypt built up a significant market share: the USA, Japan and Australia. Egypt attracts insignificant numbers of tourists from Canada, Latin America and sub-Saharan Africa. Tourism arrivals are evenly spaced between the summer months (April–September) and winter months (October–March). The tourist traffic in the past few years is summarized by main region of origin in Table 8.1.

The majority of foreign tourists are on leisure trips rather than business, and an estimated 60 per cent are attracted to the country by the historical attractions of the Nile Valley. In general, most visitors take one of a variety of trips along the Nile valley between Cairo, Luxor, Aswan and Abu Simbel. The traditional 'cultural' visits, however, do not bring about much repeat tourism.

Recent government policy has been oriented towards spreading tourism more evenly throughout the country rather than continuing to concentrate heavily on the traditional historic sites of the Nile Valley. Implicit in the policy is diversification, by providing new tourism products which would allow Egypt to appeal to a much wider overseas market than was possible in the past. The coastline along the Red Sea and the Gulf of Aqaba are being promoted for beach tourism. In recent years there has been a rapid expansion in hotels and tourism facilities along the Red Sea coast south of Hurghada and along the coast of the Gulf of Aqaba. This has begun to attract a number of visitors for sea coast tourism as well as those whose main pursuit is diving. This is a relatively specialized though potentially lucrative market, and rapid rates of growth have been achieved. There are assurances that the beach market in these areas will be able to compete with comparable Mediterranean destinations such as Morocco, Tunisia, Greece and

Table 8.1 Egypt: trends in tourism by origin

Tourists arrivals (000) from	1992	1993	1994	1995	% change 92–95
Middle East	825	767	819	741	−17.0
Africa	204	187	153	180	−11.7
Americas	224	187	182	229	2.2
Europe	1665	1205	1248	1811	8.7
Asia-Pacific	187	157	180	219	17.1
Other	121	223	301	209	72.7
Total	3206	2507	2581	3133	−2.3
Tourist nights (millions)					
Middle East	6.99	4.98	5.71	5.78	−17.8
Africa	1.49	1.07	1.07	1.05	−29.0
Americas	1.31	1.00	1.00	1.6	11.4
Europe	10.97	7.20	6.65	10.74	−2.18
Asia-Pacific	1.07	0.80	0.96	1.98	85.0
Other	0.91	1.2	0.28	0.23	−74.7
Total	21.83	15.08	15.43	20.46	−6.2

Totals may not add up due to rounding
Source: Egyptian Ministry of Tourism

Turkey. However, Egypt has an advantage over these destinations in terms of a longer season, lasting for almost eight months of the year.

Another specialized segment of the tourism market which has recently acquired prominence in Cairo is that associated with conference and incentive travel. There are a number of possible venues in Cairo which can accommodate large conferences. Statistics suggest that there may be some 250,000 visitors per annum in this segment of the market.

Until the Gulf Crisis intervened, tourism in Egypt was enjoying a period of exceptional expansion. This was due partly to the liberalization policy and partly to strong economic growth during the 1980s, particularly in the OECD countries. As a result of the Gulf Crisis, growth has steadily slowed and tourism arrivals in 1990 were close to the 1989 levels of 2.5 million arrivals and about 20 million nights: roughly half a million short of the projected arrivals. Like other tourism destinations, Egypt is vulnerable to political instability in the region and has suffered setbacks following the Achille Lauro affair in 1985, the police riots in 1986 and the US bombing of Libya shortly afterwards. However, Egypt has shown resilience, and after the Gulf Crisis and a number of terrorist events, signs of recovery are apparent with tourist arrivals during 1995 reaching more than 3.1 million.

Institutional setting

The Ministry of Tourism (MOT) is the main authority dealing with tourism in Egypt. The Ministry is organized into four major functional parts dealing with planning and development, regulation of touristic services, administration, and financial and legal affairs. Like most other ministries, MOT suffers from overstaffing and inadequate technical capability. The Ministry is being streamlined to strengthen its technical expertise in support of a private sector-led tourism development strategy, to be competitive with neighbouring countries and to protect the country's unique cultural and natural resources. The first step has been the creation of the Tourism Development Authority (TDA) in September 1991, which draws principally on private sector expertise to assist MOT in guiding and promoting increased private sector investments in the sector. These changes are expected to provide the sector with a stronger institutional framework for coherent, private sector-oriented and environmentally sound tourism development.

The Ministry also oversees the following public sector organizations: (a) Egyptian General Authority for Promotion of Tourism; (b) Public Authority for Conference Centres; and (c) Tourism Development Authority. Under Law 203, the public sector Tourism Authority, which was under the supervision of MOT in the past, became a holding company for the sector and consists of the following affiliated companies:

1 Egyptian General Organization for Tourism and Hotels (EGOTH)
2 Misr Travel Company
3 Egyptian Hotels Company
4 Misr Hotels
5 Grand Hotels of Egypt

Prevailing conditions in the sector

The cabinet and MOT have taken important steps both the at macro- and sectoral levels towards the liberalization and deregulation of the tourism sector. These include the liberalization of exchange rates, which brought about an immediate response in terms of the increased number of tourists visiting Egypt; the easing of restrictions on chartered flights; and inviting private airlines to introduce international scheduled services on routes not served by Egypt Air, Egypt's flag carrier. An institutional framework has been set up to facilitate private sector participation. Notwithstanding these reforms, the sector still faces minor constraints which impede the implementation of a private sector-led development strategy for the industry.

Public and private sector partnerships   The tourism sector has historically depended on budgetary allocation for its investments. A small share of its investment programme is financed from internally generated revenues and the rest is allocated under the five-year budgetary process. These allocations have usually been either insufficient to maintain the growth of complementary infrastructure in the sector, or were provided in an untimely fashion which resulted in serious delays in the completion of major projects. Planned investment for the tourism sector during the second plan (1986/7–1991/2) was about Eg.p828 million, about Eg.p408 million of which would be from the public sector, while about Eg.p420 million was assigned to the private sector. The sector has always been attractive to private investors, and private investments have extended to hotels, holiday villages, Nile cruises, travel agencies and tourist investment companies. Actual investments totalled more than four times the planned investments by the private sector. Since 1992 more than three billion Egyptian pounds have been invested in tourist development projects by private investors.

At present, Egypt's tourist accommodation capacity has reached 150,000 beds in about 760 hotels and a general average rate of occupancy in 1995 was 67 per cent in rooms and 49 per cent in beds. Egypt's accommodation capacity is expected to rise to 250,000 beds (130,000 rooms) by the end of 1999. The private sector is being encouraged through various incentives to introduce area infrastructure in new tourist regions by large investment companies. An example would be the Sahl Hashish area (thirty-two square kilometres) on the Red Sea south of Hurghada, which is being developed by the newly formed 'Egyptian Resorts Development Corporation' with a subscribed capital of US$200 million. This company would qualify to benefit from a World Bank Loan to the government to build tourism infrastructure in the Red Sea region. Another private sector company was given the franchise to build Egypt's first private airport in Mersa Alam on the Red Sea.

Because the tourist sector, despite its contribution to the economy, needed the database necessary for indicative planning, this has already been remedied and a computerized data system is already installed and operated through the Cabinet Centre of Information and Decision Making Support.

Recent measures to ease constraints

The Council of Ministers, in two successive meetings on 7 February and 7 March 1996 issued several decisions aiming at easing some of the existing constraints, among which were the following:

- Giving the Ministry of Tourism the right to issue a certificate of exemption from custom dues for equipment and machinery (including spare parts) necessary for tourism investment projects
- Facilitating procedures for the licensing of tourism investment projects and their renewals
- Approval of air charter traffic from all foreign departing points to all Egyptian airports except Cairo. However, charter planes carrying tourist groups may land at Cairo airport on their way back from the Egyptian landing point to their departing points abroad
- Allowing charter planes to land unconditionally in Cairo airport if their trips originate from foreign cities that are not served by Egypt Air
- Facilitating the foundation of private sector companies to develop a tourist infrastructure in various tourist areas
- Accelerating procedures to issue construction permits for hotels and other tourist establishments within city boundaries
- Reduction of passenger ship docking fees in Egyptian ports by 75 per cent
- Facilitating the creation of marinas on the Red Sea and the Mediterranean to enhance yacht tourism
- Mounting an integrated plan for the promotion of curative tourism in areas with healing characteristics such as Helwan, Aswan, Siwa, etc.
- Developing an integrated project for tourism health insurance and an air ambulance system in all Egyptian tourist regions and cities
- Providing and implementing an integral national programme for raising popular awareness of tourism and coordinating efforts between respective government departments and the media
- Preparing draft bills to amend a number of legislative provisions, e.g. Law no. 1/1973 on tourist and hotels establishments as well as the Law no. 85/1968 on tourist chambers and their federation, which require some changes to suit the new tourism developments

Tourism strategy

Despite the fact that Egypt has not yet formulated a binding national tourism policy, deliberations between the government and World Bank resulted in agreement to a number of goals, namely:

1 changing the role of the public sector from that of owner/operator to planner/regulator and promoter/facilitator;

2 deregulating the industry to allow the private sector to operate freely in a competitive environment;

3 protecting and conserving the unique cultural and natural resources in the tourism areas; and

4 promoting a larger role for the private sector in the design, financing, implementation, ownership and operations of tourism facilities.

These goals are expected to substantially increase the role of the private sector in the development of new facilities and operation of existing public sector facilities acquired through privatization and divestiture, or leasing arrangements. Private sector investments, however, have tended to concentrate on the narrow corridor of the Nile Valley and Cairo, where infrastructure is already developed and where tourism demand has been growing substantially. Attractive but remote sites, suitable for leisure tourism, have to be developed with the residual resources, resulting in unattractive facilities and inadequate infrastructure. Tourist facilities along the Red Sea coast have evolved in response to demands and in the absence of long-term land-use plans. Today, Hurghada, the first site a few years ago to be developed for tourists and vacationers, stands as an example of unguided development, inadequate infrastructure and virtually absent environmental management.

The government's approach to stemming this undesirable process of the development of Red Sea coastal areas was to give responsibility by decree to the Ministry of Tourism for planning, developing and monitoring tourism development in the remaining coastal areas and some of the areas in the Nile Valley. The Ministry's approach to development of these areas was the initiation of extensive land-use plans undertaken with the assistance of international consulting firms, and to promote private sector investments in accordance with these plans. Initially MOT's approach was to assume responsibility for infrastructure and let the private sector develop the superstructure. This approach implied that MOT would secure loans and credits and use its own resources to develop infrastructure on a large scale, with the hope that development would allow the private sector to mobilize the funds needed for superstructure. In this case, the public sector would bear all risks associated with the development of infrastructure. Provision of publicly financed infrastructure without parallel development of superstructure would result in land speculation, allowing the private sector to capture the rent associated with the ownership of land where services were developed by the public sector.

A different approach is needed if MOT's strategy for the Red Sea and Sinai areas is to succeed. There is a need for an integrated development of infrastructure and superstructure in an environmentally sound manner, and for private sector investments to be based on the principles of project finance under limited recourse, where lenders and investors assume well specified risks and rely on the project's ability to generate revenues needed for securing the loan and providing the returns. Under this concept, the private sector would assume all the commercial risks (i.e. market, viability, design, finance, construction and operation). This approach has the advantage of subjecting projects to the scrutiny of the financial community, and avoiding the need for direct intervention by MOT in project design and implementation, thus reducing the probability of resource misallocation. Furthermore, a gradual approach to infrastructure development by the private sector would allow the public sector to postpone major investments in trunk infrastructure until they were economically or environmentally justified. The Ministry of Tourism has solicited international assistance in spearheading a private sector-led strategy for the development of infrastructure in the Red Sea and Sinai areas. This is already under way.

At present, Egypt is conscious of the fact that planning for sustainable tourism is the key towards a successful tourism development strategy. An environmental impact assessment study has become the *sine qua non* of any tourism development project, without which approval cannot be granted.

SUMMARY

Sustainable tourism in any destination is the end-product that the destination should strive to achieve. This overall goal is not easy to reach, particularly in developing countries. The reason for such difficulty resides in the cost involved and in the lack of patience to pursue a balanced and selective tourism development because of the need to balance trading accounts and earn more foreign exchange. Moreover, there are many other challenges and difficulties facing world tourism that have to be tackled head on.

Tourism is a social phenomenon that reflects a society's values. National tourism policies influence a society but are also influenced by it in turn. At a time when it appears that the perception of what is desirable is widening, the concept of development and the implementation of a decision making process in pursuit of it are changing. As environmental protection and involvement of host communities − two of the basic tools of sustainability − become more significant, the requirements of tourism policy should be observed and strictly implemented. Economic considerations in developing countries should not override major social values. If tourism is to develop and each country is to reap more benefits from it, a government cannot confine itself to promoting the national heritage; it must also be able to protect and safeguard that heritage, to present it to its full advantage, and ensure that its own citizens benefit as much from it as do foreign visitors.

In the face of fierce competition between tourist destinations, developing countries should consider economies of scale and their respective carrying capacities. Tourist attractions are not the only criteria in winning this competition. The complementarity between attractions, facilities, amenities, quality services, accessibility, environmental quality and hygiene, is of prime importance.

Policy making inevitably becomes more complex when qualitative factors have to prevail. The value judgements that underpin final policy choices have to be based on a large number of factors. This requires specialized expertise, and the respective tourism authorities need to be able to make use of such expertise through bilateral and multilateral cooperation.

REFERENCES

De Kadt, E. (1979) *Tourism: Passport to Development*, Oxford: Oxford University Press.

——(1992) 'Making the alternative sustainable in tourism alternatives', in R. Butler and D. Pearce (eds) *Tourism Alternatives*, Philadelphia: University of Pennsylvania Press, 47–55.

Doggart, C. and Doggart, N. (1996) 'Environmental impacts of tourism in developing countries', *Travel and Tourism Analyst*, 2: 71–86.

Erbes, R. (1973) *Tourism and the Economy of Developing Countries*, Paris: OECD.

Erlet Cater (1993) 'Ecotourism in the Third World: problems for sustainable development', *Tourism Management*, 14, 2, 85–90.

Hong, E. (1985) *The Third World While It Lasts*, Penang: Consumers' Association of Penang.

Jenkins, C. (1992) 'Tourism in Third World development: fact or fiction?', inaugural lecture, University of Strathclyde.

Lea, J. (1988) *Tourism and Development in the Third World*, London: Routledge.

Murphy, P. (1985), *Tourism: A Community Approach*, New York: Methuen.

Pigram, J. (1995) 'Resource constraints on tourism: water resouces and sustainability', in R. Butler and D. Pearce (eds) *Change in Tourism*, London: Routledge, 208–28.

Poon, A. (1993) *Tourism, Technology and Competitive Strategies*, Wallingford: CAB International.

Schumacher, E. F. (1973) *Small is Beautiful*, London: Blond & Briggs.

Sessa, A. (1976) 'The tourism policy', *Annals of Tourism Research*, 3(5): 25–33.

Theuns, L. (1987) 'Government action and tourism sector development in the Third World: a planning approach', *The Tourist Review*, 2, 14.

UNCTAD (1973) *Elements of State Policy in Developing Countries*, New York: United Nations.

United Nations Organisation (1992) *Agenda 21, Rio Declaration*, Rio de Janeiro, Brazil: United Nations.

Young, G. (1973) *Tourism: Blessing or Blight?*, Harmondsworth: Penguin.

Wahab, S. (1974) *Elements of State Policy on Tourism*, Turin: Italgraphica.

——(1975) *Tourism Management*, London: Tourism International Press.

——(1992), 'Government's role in strategic planning for tourism', in VNR's *Encyclopaedia of Hospitality and Tourism*, New York: van Nostrand Reinhold, 746–53.

Williams, A. M. and Shaw, G. (1988) *Tourism and Economic Development*, London: Belhaven Press.

9

CHALLENGES TO TOURISM IN INDUSTRIALIZED NATIONS

Stephen L. J. Smith

Tourism is important both as a domestic industry and an export industry in industrialized nations, and will continue to grow in the foreseeable future. Some nations such as Australia, Canada and Spain have even identified the tourism industry as of strategic importance in their economic future and have initiated policies to promote the growth of tourism even more. However, the sustainable growth of tourism in industrialized nations is confronted by challenges that affect the ability of the industry to receive favourable policy attention from governments, that limit the ability of tourism business managers to make informed decisions and that influence the availability and quality of tourism products for consumers. The degree to which these challenges can be met will strongly influence the degree to which tourism can function as a sustainable industry.

This chapter explores some of these challenges. While tourism also presents challenges to societies, the focus here is on challenges to tourism. Thus the orientation might be described as supply-side in that the chapter examines challenges faced by firms, organizations, and the industry at large. The challenges are grouped into three categories: conceptual challenges, analytical challenges, and delivery challenges. The assignment of challenges to any category is somewhat arbitrary because most challenges represent a combination of, for example, conceptual issues and analytical issues, or of analytical issues and delivery issues. Each challenge is summarized and the implications briefly described. The sections also outline the beginning of a solution to each challenge.

INTRODUCTION

Tourism scholarship is replete with discussions of the social impacts of tourism. The challenges posed by tourism to the integrity of social structures, the dynamics of community life and the quality of personal and family lives in host communities are frequent topics of research. The relationship between the industry and the destination community is, of course, two-way. Social forces affect tourism as surely as tourism affects society. This chapter

147

explores some of the challenges created by social structures and values for the industry. A better understanding of these challenges is important because they affect the ability of the industry to receive favourable policy attention from governments, they limit the ability of tourism businesses to make informed management decisions, and they affect the availability and quality of tourism products for consumers. The degree to which tourism can meet these challenges will determine to a significant degree whether tourism can function as a sustainable industry.

This chapter is selective in its coverage. There are many problems confronting the industry, such as economic recession, crime and security, and global warming. Many problems, however, are limited to specific regions or are long-term. Only fundamental, immediate industry-wide issues common to industrialized nations are discussed here. These are grouped into three categories: conceptual challenges, analytical challenges and delivery challenges. The assignment of challenges to any category is somewhat arbitrary because most represent a combination of, for example, conceptual and analytical issues, or of analytical and delivery issues.

Each challenge is summarized and its implications briefly described. The sections conclude with a brief statement of steps that might lead to a strategy to deal with each challenge. Finally, it should be noted that while the focus of this chapter is on industrialized nations, the challenges identified often affect tourism in developing nations as well.

CONCEPTUAL CHALLENGES

- Tourism is a non-traditional industry
- The tourism product depends on the effective integration of many different commodities
- Tourism data too often are incompatible, inconsistent, and not credible

These interrelated issues are arguably the greatest challenges facing tourism. A lack of understanding of the nature of the tourism industry and the tourism product even within the industry itself has plagued tourism analysts, policy makers and leaders for years. An illustration of this problem can be seen in the following quotation:

> trade in tourism services may be thought of as arising due to demand by itinerant or 'footloose' consumers. *It* [tourism] *is not an 'industry' in the conventional sense as there is no single production process, no homogeneous product and no locationally confined market.*
>
> (Tucker and Sundberg 1988: 145; my emphasis)

In one sense, Tucker and Sundberg are correct: tourism is not a conventional industry. However, their implication seems to be that tourism is not an industry in any sense. Rather, in their view, tourism is a composite of a

148

number of industries: transportation, attractions, accommodation, food services, retail trade, and so on. Many of the commodities produced by these tourism industries also are purchased by non-tourists. For example, airline passenger service and hotel rooms are consumed by people shifting their permanent residence (and thus are not tourists); meals in restaurants are consumed by workers and shoppers; museum exhibits are attended by local school groups; and crafts are purchased by local collectors. Further, tourists also purchase non-tourism commodities such as groceries, newspapers, clothing and healthcare products. If tourism is an industry, it obviously is not a conventional one.

An industry is defined by the product it produces. This is the core of the problem some analysts have with considering tourism as an industry. They argue tourism is not an industry because it lacks a generic product. Smith (1994), in contrast, argues that both a generic product and a generic production process exist. However, just as tourism is not a conventional industry, the tourism product is not a conventional product. The tourism product is the complete travel experience: a composite of a number of different commodities such as transportation, food and beverages, accommodation and attractions that provides the consumer with the complete set of services necessary to support the travel experience. The individual commodities typically are purchased from different businesses and are combined into the final product. This is usually done by the individual consumer, although an all-inclusive tour is the creation of an integrated product by a tour wholesaler.

One result of the complex mix of commodities that constitute the generic tourism product, as well as the diverse purchasing patterns of tourism and non-tourism commodities by tourists and non-tourists, is that measuring the economic effects of tourism – GDP contributions, tax revenues generated, jobs created, payrolls supported, imports and exports produced – in a way that is consistent with those used for conventional industries, is difficult. In turn, accurate and credible comparisons between tourism and other industries are also difficult. Without measures of the size of the tourism industry based on methods consistent with other industries, it is too easy for policy analysts and elected officials to dismiss any claim about the importance of tourism. This lack of reliable and accurate data on the economic performance of the industry makes scientific monitoring of the industry's health and sustainability impossible.

Another aspect of these challenges, especially the lack of credible and consistent data, is that it was not until the 1991 Ottawa Conference (The World Tourism Organisation's International Conference on Travel and Tourism Statistics) that international agreement was reached on basic definitional and statistical conventions related to domestic travel. Although guidelines for defining and measuring international travel have been in place since 1937, there were no agreements regarding domestic travel. Most countries have poor data on their domestic industry. In fact only Australia and

Canada among industrialized nations have relatively long-term, comprehensive statistics on domestic travel patterns. The visibility and credibility of an industry are directly related to the availability of credible data on economic activity associated with that industry. To put this as an aphorism: if you cannot count it, it does not count.

Of course, one must count accurately and consistently. The well known claims of the World Travel and Tourism Council (WTTC 1995) about tourism being the world's largest industry provide an example of how not to measure the size of an industry. The Council not only estimates the value of tourism products sold to consumers, they include measures that substantially exaggerate their final figures. For example, their estimates include capital investments in infrastructure such as highway construction and telecommunications that may be used by tourists but cannot legitimately be considered to be part of the tourism industry. The Council includes expenditures for all travel, not just that which meets WTO's definition of 'tourism'. They also rely heavily on secondary data sources, even when more reliable primary sources are available. For example, in making estimates for Canada, the Council uses US data, including the US input-output matrix as the basis for their estimates. The result of WTTC's approach, in the case of Canadian tourism, is an estimate of the 'size of the tourism industry' that is perhaps three times larger than actual! A detailed critique of the Council's methods can be found in Wilton (1996).

Part of the solution to these problems is the tourism satellite account (WTO 1994; Meis and Lapierre 1995; Smith 1995a). A satellite account is an information system that is a subset or 'satellite' of a nation's national accounts. Using an input-output model of commodity flows among all industrial sectors in a nation's economy, a satellite account accurately describes the level of commodities demanded by and supplied to tourists in all sectors of a national economy. Problems of double counting and idiosyncratic definitions of 'industry', such as those that plague WTTC's estimates, are avoided.

In order to develop a satellite account, a nation must have a system to produce reliable and accurate data on a continuing basis. A satellite account is only as good as the data available for it. Thus a satellite account can help provide the motivation and direction needed to create a national tourism data collection system that will serve a variety of purposes.

A tourism satellite account collects, orders, and interrelates statistics describing all significant and measurable aspects of tourism within a framework that organizes tourism data according to real-world relationships. It provides a measure of the economic magnitude of tourism in a form that is disciplined, balanced, and directly comparable to measures of traditional industries. Beyond this, it demonstrates at a macro level how various commodities are combined into the integrated experience that is the tourism product. However, as useful as satellite accounts are for economic measurement, they do not provide data related to environmental aspects of sustainability. Work on these information tools is being conducted under the aegis of the International Working Group on Indicators of Sustainable Development (1993). A review of their draft indicators is summarized by Smith (1995b: 295–300).

The 'hospitality industry' is an oxymoron

Hospitable relationships between visitors and hosts is a prerequisite for a sustainable tourism industry. Hospitality to visitors also is an ancient tradition. Muhlmann (1932), in a dated but still useful review of the concept, notes that a number of forces shaped pre-industrial societies' traditions of hospitality to travellers. Travellers were strangers and often were feared because they were unfamiliar and sometimes even suspected of possessing magical powers. A friendly reception might help avert a dangerous confrontation. As Muhlmann puts it, 'one protects the stranger in order to be protected from him' (1932: 463). Strangers also brought trading opportunities and news from neighbouring lands, so a friendly reception might produce some benefit for the hosts. To paraphrase Muhlmann's formula, one gives gifts to strangers in order to get gifts from them. Over time, these motives were transformed from acts of reciprocity to ethical or religious obligations. The point here is that hospitality is fundamentally an organic, personal relationship between a host and a guest.

Since the Industrial Revolution, hospitality – particularly in the form of providing accommodation, and food and drink – became a set of commodities. Rather than being a spontaneous and personal gesture from one individual to another, accommodation and food and beverage services became something a traveller could demand through a cash payment. The terms of this commercial exchange became governed not by traditions or religion, but by legal codes and third parties such as law courts.

This is a tourism issue for the following reason. Travellers still believe implicitly in the ideal of being treated as a valued guest. They want not only competent service but 'service with a smile', a warm welcome, personal attention – service that 'exceeds expectations'. Educators as well as managers of hospitality firms in the tourism industry understand this and attempt to inculcate these values into their students or employees. However, many employees perceive their role not as that of a gracious host but as an underpaid, overworked worker in a dead-end job. The result can be a significant gap between what the visitor believes is their right as an honoured guest and what the service-delivery person intends to offer. There is often a fundamental difference in the perception of the nature of the product to be delivered. Beyond a basic specification of any given tourism product, e.g. a half-day sightseeing tour or a lobster dinner for a specified price, tourism products as experiences remain undefined and often imbued with divergent expectations by producer and consumer.

This issue needs to be addressed on both a practical and a conceptual level. On the practical level, the training, self-perception and conditions of employment of tourism workers need to be improved. This is addressed in greater detail in the section below on delivery challenges. Conceptually, scholars need to examine critically the nature of hospitality, gifts and exchanges in industrialized societies (Burgess 1982; Wood 1994). Such an analysis may lead eventually to a better understanding of the dynamics of the exchange process governing hospitality services from both the provider's and the recipient's viewpoint. It may also help ensure that tourism produces opportunities that are satisfying and appropriate for workers in the community seeking employment.

Confusing tourism as a means versus an end

Governments are interested in tourism for the economic benefits it can provide. Conservation groups want tourism to pay for environmental protection. Arts organizations use tourism to justify public support of their programming. Aboriginal cultures view tourism as a strategy to improve the economic health of their communities. Many groups perceive tourism as a means to serve their own ends. This perception is understandable and can, within limits, provide the base for successful partnerships between tourism organizations and non-tourism groups. However, it can also lead to frustration and disillusionment. Problems, when they occur, stem ultimately from forgetting the nature of tourism.

Tourism, from a demand perspective, is the set of activities engaged in by individuals temporarily away from their usual environment for any reason other than the pursuit of remuneration from within the place visited. From a supply perspective, tourism is the collection of businesses and other organizations that facilitate the activities of people temporarily away from their usual environment. To the extent that the goals of visitors (such as a satisfactory leisure experience) and the goals of tourism businesses (such as market share) can be met through cooperation with non-tourism organizations, partnerships can be successful. But a convergence of goals cannot be taken for granted, especially if a third party (other than the tourist and the tourist business) with its own agenda gets involved.

Governments sometimes view tourism as an easy target for tax increases. They incorrectly perceive visitors as being price-insensitive or as a politically safe target for excessive taxation because tourists do not vote in the jurisdiction that sets the tax. Conservation agencies may impose restrictive guidelines on visitors to environmentally sensitive areas and yet still expect sufficient numbers to arrive and to spend sufficient money to pay for environmental protection. Arts organizations often naively believe visitors will 'just come' without making any investment in marketing. In fact, the same arts organizations that invoke tourism to justify receiving public grants

152

sometimes disdain presenting themselves as tourism attractions. Aboriginal groups may conduct business in ways that are traditional in their culture but not consistent with the expectations of travel service wholesalers and retailers.

Meeting this challenge requires that tourism organizations understand that their goals and those of potential partners are not necessarily compatible despite the initial enthusiasm some groups display when they 'discover' tourism. Tourism organizations must make their goals clear and seek to understand the goals of potential partners before commitments are made or joint projects initiated. The lack of coherence between the goals of tourism and the uses to which non-tourism organizations wish to put tourism does not necessarily mean that one is right and the other is wrong. At a minimum, it means that some compromise or trade-off might be necessary. This leads directly to the first analytical challenge facing the industry.

ANALYTICAL CHALLENGES

- Tourism planning and development requires trade-offs
- Tourism planning and development often involves conflicting or incommensurate objectives

The necessity of trade-offs, of allocating scarce resources among users whose competing demands for those resources exceed their supply, is a familiar problem. Resources are limited, whether beachfront property, potable water, money or time. The challenge for tourism is not so much the need to make decisions about resource allocation as the development of analytical methods to assist in allocation. This statement is not intended to imply that making trade-offs would be easy if only one had the right analytical tools. Analytical tools will never replace human judgement in tourism planning and development. Further, special interest groups often resist any trade-off if it means their own agendas are not fulfilled. But beyond these issues, the field still lacks adequate tools to guide trade-off decisions.

Trade-offs occur in two contexts. The first is internal, or the need to allocate resources within the context of a single organization. The second is external, or the search for a balance among potentially conflicting goals of a tourism organization and other groups. Internal allocations may be guided by the principle of maximizing net returns. For example, a resort developer will, in theory, allocate various amounts of land to accommodation, golf courses, tennis courts, commercial shops, support services, maintenance facilities, parking and other uses in such a way as to create the physical environment that will most likely maximize profits or market share, or achieve another business goal.

Complications in the application of this principle arise from several sources. There usually is uncertainty about the exact causal relationship between the percentage of space allotted to any use and overall cashflow. The volatility of consumer tastes increases the uncertainty about the optimal mix of land uses and resource allocation. Environmental constraints in the form of soil instability, terrain, flood hazards, or the desirability or regulatory necessity of protecting environmentally sensitive or significant parcels, may preclude a financially optimal solution. Governments' use of economic instruments to achieve certain policy goals can also complicate rational resource management decisions. Instruments such as 'user/polluter pays' schemes, restrictions on numbers of users and other capacity control techniques, and various educational techniques and codes of conduct, have a place in public policy, but their use can create problems or even worsen the situation they were designed to correct (Forsyth *et al.* 1995).

Although allocation tools have been developed to assist in the optimal use of resources, they have limited value for many tourism businesses. For example, linear programming allows an analyst to identify quantitatively how the level of one or more outputs varies as the level of inputs changes. Linear programming works well with relatively simple allocation problems such as determining the balance of irrigation water, fertilizer, labour and land assigned to a specified crop that will maximize yield. The solution to this type of problem involves known and measurable relationships between inputs such as water and outputs such as crop yield. Further, all inputs and outputs ultimately can be compared by converting them to a common unit of measurement such as financial value.

While tourism businesses also utilize inputs and produce outputs that are measurable in financial terms, the relationships between, for example, money spent on staff training (as an input) and revenue generation (as an output) cannot be specified as a precise function. For example, while it is customary for tourism officials, business leaders and educators to pay lip service to the importance of staff training, there have been few studies measuring the rate of return on investment in staff training. Instead, industry leaders rely on personal experience or anecdotal evidence to make decisions about staff training. The problem of being able to specify objectively a causal linkage among quantifiable inputs and outputs is only part of the challenge. Another part of the challenge is that some factors involved in tourism production such as staff attitudes, changes in tastes and expectations of consumers, and governmental regulations, are non-quantifiable.

Part of the solution for many resource allocation problems will be the development of guidelines based on the experience of established firms. Average performance ratios, such as the mean housekeeping cost per room for hotels of a specified type, or the percentage of the operating budget spent on advertising by attractions of a specified type, can provide some benchmarks for managers. Best-practice business profiles could also provide guidance. It must be acknowledged, of course, that developing detailed guidelines is difficult because of the reluctance of individual businesses to release proprietary financial data for public dissemination. The use of averages derived from multiple firms' data or of techniques to disguise the identity of actual cases may help overcome this problem.

An even more complicated challenge is balancing competing demands for scarce resources in a public context. Not only is it difficult to quantify relationships between inputs and outputs, but the demands, expectations, costs, and benefits associated with the use of public resources involve multiple parties and interest groups. A golf course provides profits for its owners and recreation for visitors but removes arable land from a region's agricultural base and may contaminate surface water as a result of the heavy use of fertilizers and pesticides. A national park protects a fragile biome but displaces private residences and closes traditional hunting and fishing grounds. The pursuit of legitimate private or public goals, whether for profit or preservation, may result in benefits and costs that affect others.

These benefits and costs are called 'externalities' because they are external to the organization involved in the project. Private sector enterprises normally ignore externalities unless forced to take them into account by governmental regulations. Public sector enterprises, however, usually must consider externalities.

Benefit-cost analysis can be used to assist in the evaluation of whether the overall public benefits outweigh overall public costs. Benefit-cost analysis, though, requires that benefits and costs be measured in monetary terms to permit rational and consistent assessments. Needless to say, not all benefits and costs can be measured in a monetary metric. Further, benefit-cost analysis typically ignores the social distribution of costs and benefits. While benefit-cost analysis indicates whether total benefits outweigh total costs, it does not address who receives the benefits, who pays the costs, or if the allocation of benefits and costs is equitable. Much more work needs to be done to develop reliable and credible techniques that will help analysts and decision makers to consistently and objectively compare the full range of benefits and costs arising from tourism development. Such refinements should include the refinement and testing of measures of sustainable tourism development, and their conceptual linking with the methods and conventions of benefit-cost analysis.

DELIVERY CHALLENGES

Many tourism employers put too little importance on human resource development

Tourism's potential for job creation is one reason why governments are interested in promoting its development. A given level of investment or revenue creates more jobs in tourism than the same level of financial input in most other economic sectors. Tourism also provides a high percentage of entry-level jobs for people entering the workforce for the first time.

As a labour-intensive industry, the quality of service provided by tourism employees profoundly affects the visitors' level of satisfaction and the probability of repeat business for tourism firms. Unfortunately, tourism jobs, especially those on the front line, often are viewed as long hours of low-paying menial labour with no prospect of advancement or a long-term career. Low employee morale, high turnover and lack of job commitment resulting from the conditions found in many tourism jobs are understandable, although regrettable.

Despite the importance of tourism as a source of job creation and work experience for people entering the workforce for the first time, and the importance of quality service as a determinant in the customer's satisfaction with the tourism product, relatively few tourism businesses invest in employee training and retention or attempt to improve the conditions of employment. The rates of employee turnover in tourism and hospitality are among the highest of any industry. Many businesses find it difficult to attract enthusiastic and committed employees. Some employers complain about the poor-quality work they receive from their staff, yet refuse to provide real incentives for good performance or to invest in training to improve morale and productivity. A common attitude among tourism managers is that if they train their employees, the employees will leave for more attractive positions, presumably with the business's competition. 'What happens if you don't train them and they stay?' is a question these managers appear not to address.

Part of the solution is to conduct return-on-investment studies on human resource development (it should be noted that some authors object to the concept of 'human resources', preferring to conceptualize labour issues in terms of 'employee relations'; see Lucas 1995 for an example). While there are substantial conceptual and methodological problems in demonstrating the link between expenditures on human resource development and a business's bottom line, such studies need to be undertaken. Work must be developed to propose conceptual models describing the causal connection between investment on training and the types of financial rewards it can eventually generate for a business, as well as on developing the analytical models that can help describe and measure this connection. These studies might be supplemented with best-practice profiles of professional development strategies that have been implemented by successful companies. This type of work has the potential to demonstrate both the financial value of enlightened human resource development strategies as well as to provide practical examples for better practice.

Another part of the solution may be the development of job performance standards and certification procedures for tourism jobs, especially those dealing directly with the public. Tourism jobs involve three domains of understanding and ability: factual knowledge, performance skills and attitude. Consider a fishing guide. The guide must know facts about fishing regulations, fish identification and preferred habitats and feeding habits of game species. Skills include boat handling and fishing techniques. Finally, the guide must have those attitudes and styles of personal interaction that will help them function more effectively.

Skilled trainers working closely with experienced guides can identify the range of all three types of abilities and then document them in such a way as to provide objective criteria for training the guide and for evaluating the guide's performance. These criteria can also provide the base for a formal examination process, including a combination of written or oral examination and on-the-job assessment. The combination of skills explicitly identified and taught through a formal training process with an examination and certification process may help to raise the level of self-esteem of workers in the industry, as well as management's appreciation of what constitutes a productive, competent employee. This strategy may in the long run lead to a greater emphasis on the training of front-line employees, as well as a greater appreciation – and presumably greater financial rewards – for employees who have been certified in their occupations.

The market has become mature and 'super-segmented'

Tourism in industrialized nations has virtually exploded since the end of World War II. The development of commercial jet air passenger services, historical increases in disposable income and discretionary time (both of which have stagnated or even reversed in recent years in some industrialized nations) and the growing demand for business travel have driven dramatic increases in tourist arrivals and receipts in industrialized nations. More people are travelling more often and spending more money than at any time in history.

Also since the end of World War II, tourism has seen a slow but accelerating growth in the range of products. For years, especially at the international level, the tourism industry offered a relatively narrow range of standardized products. The term 'international hotel' implied certain types of amenities, levels of service and even location. Resorts could generally be graded into four or five different levels of quality and price, and offering much the same set of activities. The great majority of commercial carriers – airlines, trains, ships and motor coaches – also offered the same types of products, with differences based on price and an associated level of service. However, as entrepreneurs have recognized the growth in tourism, they have sought ways of offering more competitive products.

A result of these two forces – an increasing number of increasingly experienced travellers and increasing competition for their business – is an unprecedented diversification of travel products. To put it more precisely, the market has become highly fragmented. Businesses, in an attempt to deal with this fragmentation, have begun to identify an ever-increasing number of consumer segments and are positioning products to cater to these segments.

Consider commercial accommodation. The types of accommodation available are now several score. They include small, intimate bed-and-breakfasts and inns that reflect local cultures; high-rise casino-hotels with thousands of rooms; all-suite or apartment hotels catering to long-term business travellers; highly standardized, low-service budget motels serving families travelling by automobile; 'day rooms' in airports providing a comfortable, secure place for in-transit passengers to doze between flights; fully serviced campgrounds with pools and other recreational facilities attracting retirees with luxurious motor homes; and unserviced wilderness campsites accessible only by hiking or canoeing. Each can be further segmented on the basis of price or location. This list could be extended, but it is sufficient to make the point: the range of accommodation options available to travellers is substantial and becoming greater. The same is true in every sector of the industry.

While this growth in diversity is good news for the consumer, who is now more likely to find the type, location, and price of product desired, it presents a substantial challenge to the industry. Tourism businesses and destinations will find it increasingly difficult to compete by positioning themselves as offering 'something for everyone'. They must identify their own competitive advantage, a market for their product, and the techniques to reach that market. Achieving this level of sophistication in product development and marketing is a special challenge for the great majority of small enterprises that make up the bulk of the tourism industry in most industrialized nations.

Another reason for 'super-segmentation' is the proliferation of generic segmentation models offered by market research consultants. In their attempt to position their own firms competitively, each consulting firm develops its own way of describing the market. The result is a bewildering array of segments and segment models that only adds to the confusion. Ironically, while many aspects of tourism suffer from a shortage of analytical tools, there is an over-supply of segmentation tools.

To return to the issue of the growing diversity of tourism products, that growth may be snowballing. As travellers discover new products, their expectations for still newer and more innovative tourism products increases. While this offers opportunities for businesses to develop new products, they face several difficult questions. Are the products being developed sustainable in the long run? What types of new, sustainable products will be accepted by the market? Is there a sufficient demand to justify the new product? Could certain new products be packaged together to make them more economically competitive? If so, how? Should older products be continued, modified or phased out? Obtaining the answers to these questions increasingly requires reliable market data and research as well as greater market intelligence about the products and experiences of competitors.

The industry is slow to respond to technological change

The explosion of information and communications technology is well known. Although the claims of commercial Internet providers such as America-Online, CompuServe and Prodigy regarding the number of their subscribers cannot be accepted because of their refusal to allow independent auditing of their numbers, there is no denying that the Internet and the Web are widely used by many individuals and organizations. CD-ROMs, laser disks, more powerful computer hardware, more sophisticated telephone systems and equipment (ranging from more powerful pageing devices to portable fax machines), multimedia technology (perhaps updated over a TCP/IP network) and a growing array of specialized software such as computer reservation systems, regionally integrated computer reservation management systems, point-of-sale software, and inventory control or flow-tracking software, are also familiar to many in the industry.

However, many tourism businesses and organizations are slow to adopt new technologies, especially small and medium-sized enterprises. Sometimes this is due to uncertainty over whether any new development is a tool or a toy. The answer, of course, is often specific to the individual organization. Some businesses will benefit from having a home page on the Web, whereas others would find a home page yielded no net returns. Further, almost anyone who has adopted a new computer system or software has experienced frustration with unanticipated problems or false promises made by the promoters of the product. Scepticism when dealing with any technology sales representative is prudent.

Even when new technology does offer real improvement, there are barriers to its adoption. In the case of computer reservations systems, for example, current systems are fundamentally layer upon layer of old code, each layer providing new features. Despite imaginative modifications, existing computer reservation systems cannot take advantage of the latest technology because they are ultimately tied to an old computer system. Replacing a system with truly new technology may involve unacceptably high costs for new hardware, new software, training and technical support.

Alternatively, reservation system providers might design user-friendly services on the Internet to permit consumers to do direct bookings. Such developments are already under way to a limited extent. One question is how system providers maintain their revenue stream if their systems are available directly to consumers, and of course there is the question of how travel agencies respond to this technological shift. The point here is not whether such change will occur, but rather that the adoption of new technology is often not just a matter of cost effectiveness, but a fog of business, social and economic concerns surrounding any technological change.

There are also personal concerns. A certain percentage of tourism managers are reluctant to accept any change regardless of how cost-effective. Some family-run businesses are led by managers who feel that just because they have been successful for years operating in one way, their way of running their business will continue to work in the future. They may cite concern over the time it will take to learn a new system or over the full costs of adopting a new technology. Fear of change – however rationalized – cannot be ignored as a factor.

As a result of fear of change, organizations may fail to take advantage of improvements that would reduce long-term costs and increase productivity. Firms that fail to adopt some technologies will eventually find themselves being bypassed by potential customers. For example, businesses that do not have fax machines will find their market disappearing because potential buyers will do business only with those firms they can reach via fax. As reservation systems become more integrated, comprehensive and an even more vital part of the distribution system, businesses not listed on those systems will become 'invisible' to the market and will slowly die.

To encourage the adoption of appropriate technology, several tactics need to be employed. First, more research is needed on the barriers to adoption of new technology; on the relative importance of potential problems – whether cost, lack of training, suspicion over the usefulness of new technology, or whatever – in specific sectors and specific communities or regions. The results of this work can then shape the responses of industry associations, governments, educational and training institutions and technology providers to overcoming the barriers. The development and distribution of best-practice case studies and the conducting of demonstration projects at typical businesses are likely to be part of the strategy that will help convince some owners and operators to adopt new tools. Workshops and seminars, offered at affordable prices and convenient locations by objective, reputable organizations in which products are demonstrated and evaluated, may also help. In the long run, tourism managers must be incorporated into research and development exercises developing new tools for the industry. (Additional general perspectives on technology and tourism may be found in Poon (1993) and Buhalis (1994).)

Investment capital for tourism development is increasingly difficult to obtain

In Canada, tourism businesses are sometimes characterized by banks as 'three months of revenue and nine months of bankruptcy'. This bleak picture refers to the highly seasonal nature of some Canadian tourism enterprises. A less-than-favourable view of tourism businesses as markets for investment capital, though, is not unique to Canada. Financial institutions in many industrialized nations fail to assess accurately the financial viability of tourism businesses because they lack information about the nature of tourism as an area of economic activity.

This ignorance translates directly into higher costs of borrowing or bars firms from obtaining investment capital at all. There are numerous problems created by the lack of capital, but one of particular concern is that firms will not be able to invest in new technologies, management tools, product improvement or staff training that will allow them to conduct their business in a way that is economically and environmentally sustainable. Instead, they will focus on coping with short-term financial problems – an approach that too readily ignores environmental and social costs.

While there are justifiable concerns about the performance of specific tourism firms in specific localities, too many governments and financial institutions still do not understand the role of tourism in regional and national economic development. They may not perceive tourism as an export industry with the potential to contribute significantly to a nation's growth and balance of payments. Further, the market place is not always capable of guiding long-term economic allocation decisions. In particular, governments' interests in improving the distribution of incomes, in promoting regional development, in job creation, and in protecting environmental or cultural resources, often are not served by a simple reliance on the mechanisms of the capital market. Tourism can play a significant role in helping governments achieve their goals in these areas if, and only if, they can obtain access to investment capital.

As growth continues in certain developing nations such as China, India and countries in Eastern Europe and South America, there will be increased global competition for capital. Industrialized nations will see increased demands for investment capital and secure rates of return as their governments attempt to rebuild ageing infrastructure such as highways and bridges and to ensure the long-term stability of pension plans. All this will result in both increased demand and greater emphasis on asset performance. Projects that do not demonstrate an attractive, secure rate of return will find it difficult to compete for capital. Not only will private money be subject to these pressures, governments are also likely to be pressed to direct their investing and loan/grant programmes to those aspects of development that are perceived as being of strategic importance to the nation.

The complexity of this challenge requires a number of long-term responses by the tourism industry. Part of the solution is to develop and disseminate credible statistics documenting the importance of tourism to a nation's gross domestic product. Industry associations or public agencies should also work with individual sectors to develop financial indicators that can be used by potential investors to assess the performance of any individual business seeking capital. Managers of tourism firms need to be better educated to increase their skills at managing costs and revenues of their businesses. This will probably mean both more skill at budget control, but also improvements in the management of their human resources and the development and marketing of competitive products. Until better risk and performance data can be made available to private financial institutions, governments might be encouraged to provide loan guarantees and other financial incentives for carefully selected tourism businesses that have difficulty competing for private capital because of inaccurate perceptions of risk (see Wanhill (1994) for a review of the role of investment incentives in tourism).

SUMMARY

This chapter has identified some fundamental challenges confronting tourism. Other challenges could be identified but these often are the results of the fundamental challenges discussed in this chapter. For example, some might suggest that the need to minimize environmental impacts and to design sustainable products should be explicitly cited as a major challenge. While one does not deny the importance of this issue, many of the difficulties in dealing with balancing environmental impacts and the needs of tourism developers are manifestations of the lack of, first, credible data on the impacts of various forms of tourism development; second, of a sound conceptual understanding of the phenomena being studied; and third, of appropriate analytical tools for identifying and assessing potential trade-offs in resource allocation. Other analysts and industry advocates argue that excessive taxation and regulations are a serious challenge. Again, this is true, and again, the challenges created by taxation and over-regulation are often manifestations of the industry's lack of credibility as an economic sector and of a lack of reliable data showing its contributions to the economy and the negative effects of excessive taxation on the health of the industry.

The challenges discussed here are classified into three discrete categories, but in fact they tend to overlap. For example, the issue of 'super-segmentation' is presented as a delivery challenge because of the implications of the diversity of products and markets for tourism producers. However, the phenomenon of ever-growing segmentation may be as much an analytical challenge as a delivery challenge. The solutions to the analytical challenges of negotiating trade-offs and dealing with incommensurate objectives requires that there be a solid conceptual understanding of the nature of tourism products and resources to start with. The lack of financing, a delivery issue, is to a large degree a result of the lack of credible data on industry performance, a conceptual challenge.

Despite the complexity and magnitude of these challenges, many analysts, planners, policy makers and industry leaders are searching for ways to meet them. Part of the strategy for solving each challenge has been addressed (admittedly in a superficial fashion) in each section of this chapter. These challenges are significant but not insurmountable. There is reason to be concerned and to be busy, but no cause for pessimism.

REFERENCES

Buhalis, D. (1994) 'Information and telecommunications technologies as a strategic tool for small and medium tourism enterprises in the contemporary business environment', in A. V. Seaton, C. L. Jenkins, R. C. Wood, P. V. C. Dieke, M. M. Bennett, L. S. MacLellan and R. Smith (eds) *Tourism: The State of the Art*, Chichester: Wiley, 254–74.

Burgess, J. (1982) 'Perspectives on gift exchange and hospitable behaviour', *International Journal of Hospitality Management*, 1: 49–57.

Forsyth, P., Dwyer, L. and Clarke, H. (1995) 'Problems in use of economic instruments to reduce adverse environmental impacts of tourism', *Tourism Economics*, 1: 265–82.

International Working Group on Indicators of Sustainable Tourism (1993) 'Indicators for the sustainable management of tourism', unpublished report, Environment Committee, World Tourism Organisation, Winnipeg, Manitoba.

Lucas, R. (1995) *Managing Employee Relations in the Hotel and Catering Industry*, London: Cassell.

Meis, S. and Lapierre, J. (1995) 'Measuring tourism's economic importance: a Canadian case study', *Travel and Tourism Analyst*, 2: 78–91.

Muhlmann, W. E. (1932) 'Hospitality', in E. R. A. Seligman (ed.) *Encyclopedia of the Social Sciences*, New York: Macmillan, 463.

Poon, A. (1993) *Tourism, Technology and Competitive Strategies*, Oxford: CAB International.

Smith, S. L. J. (1994) 'The tourism product', *Annals of Tourism Research*, 21: 582–95.

——(1995a) 'The tourism satellite account: perspectives of Canadian tourism associations and organizations', *Tourism Economics*, 1: 225–44.

——(1995b) *Tourism Analysis*, 2nd edn, Harlow: Longman.

Tucker, K. and Sundberg, M. (1988) *International Trading in Services*, London: Routledge.

Wanhill, S. R. C. (1994) 'Evaluating the worth of investment incentives for tourism development', *Journal of Travel Research*, 33 (3): 33–49.

Wilton, D. A. (1996) *The Economic Significance of Tourism: Comparing Canada and WTTC Estimates*, Ottawa: Canadian Tourism Commission.

Wood, R. C. (1994) 'Some theoretical perspectives on hospitality', in A. V. Seaton *et al.* (eds) *Tourism: The State of the Art*, Chichester: Wiley, 737–42.

World Tourism Organisation (1995) 'A satellite account for tourism', unpublished working paper, Madrid: World Tourism Organisation.

World Travel and Tourism Council (1995) *Travel and Tourism's Economic Perspective: A Special Report for the World Travel and Tourism Council*, Brussels: WTTC.

10

IMPLEMENTING SUSTAINABLE TOURISM DEVELOPMENT THROUGH CITIZEN PARTICIPATION IN THE PLANNING PROCESS

Craig Marien and Abraham Pizam

This paper examines various techniques for citizen participation in tourism development. The techniques are categorized based on their effectiveness in achieving certain participatory objectives. Encouraging citizens to participate requires the opening of power distribution channels and legitimizing tourism for the citizens. The implications of power and legitimacy are studied, and some suggestions for overcoming problems associated with power distribution and legitimizing participation are presented.

INTRODUCTION

Increased income, jobs and development are those attributes of tourism that many communities seek. Tourism, however, is not without its negative side-effects. Tourism can lead to pollution, congestion, noise, crime and visual deterioration to the community. These effects reduce the quality of life for residents and the pleasantness of the experience for tourists. Consequently this may lead to a decline in tourism receipts and arrivals, which in turn affects the health of the local economy. Today there is a consensus among researchers and practitioners alike that growth and development *per se* are not necessarily desirable, and that sustainable development is the only type of development that 'can meet the needs of the present without compromising the needs of future generations' (World Commission on Environment and Development 1987). Rees' (1989) definition of sustainable development is most appropriate not only for tourism development but also highlights the importance of citizens' involvement in tourism planning. According to Rees:

> Sustainable development is positive socioeconomic change that does not undermine the ecological and social systems upon which community and society are dependent. Its successful implementation requires integrated planning, and social learning processes; its political

viability depends on the *full support of the people it affects* through their governments, their social institutions, and their private activities.

(Rees 1989: 13; original emphasis)

As is evident from the above, sustainable tourism cannot be successfully implemented without the direct support and involvement of those who are affected by it. Therefore, evaluating a community's sensitivity to tourism development is the first step in planning for sustained tourism development. Evaluating a community's sensitivity is not an ephemeral event, it is an ongoing process. Communities are constantly evolving and so do tourism developments and their associated impacts. Therefore, devising means to allow for citizens' involvement in the process of tourism planning, and encouraging citizens to participate actively in this process, is of primary importance for sustainable tourism development.

CITIZEN PARTICIPATION TECHNIQUES

Many techniques are available for citizen participation in the tourism planning process. It is important that the techniques chosen be based primarily on the participation objectives. Many participation programmes have resulted in failure because the wrong techniques were implemented. Participation techniques can be divided into two categories based on the objectives being sought: *administrative objectives* or *citizens' objectives*.

Based on *administrative objectives*, citizen participation in tourism planning is a means of improving citizen trust and confidence in the government. The goal is to increase the likelihood that the citizens will cooperate with officials by working within the system rather than against it. Developing a greater understanding and cooperation between citizen groups and government is more cost-effective than conflict resolution and repairing damage caused by protests of angry citizens who feel that they have been negatively affected by tourism developments. Information exchange, education and support building are participation techniques that promote administrative objectives.

To meet *citizens' objectives*, citizen participation is a way for 'governments to respond better to the citizens' values and refrain from the erratic, insensitive, or oppressive exercise of power' (Arnstein 1969). Citizen participation to achieve this end is in the form of either decision-making supplements or representational input.

The *administrative objectives* techniques are those which best satisfy the local government's needs for public participation in tourism planning, while the *citizens' objectives* techniques are those that best satisfy the citizens' needs for taking an active part in the tourism development process. The best participation programmes strike a balance between administrative and citizen expectations for participation.

Administrative-oriented techniques

Administrative techniques are consultative in nature. The focus of these techniques in tourism planning is to build a consensus between government officials and the public at large on planning policies as related to tourism development. Since communities have many genuine conflicts of values and interests, there is a constant need to provide a forum for airing these differences. Even the best tourism planners cannot be expected always to know what is best for everyone. Consultation therefore is a means of opening public debate and stimulating further citizen involvement (Burton 1980). It is important to note that in most cases, local authorities have no formal obligation to incorporate their citizens' advice into any of their planning decisions. Hence consultative participation has often been referred to as tokenism. The powerholders still retain the right to decide, while the citizens have the opportunity to express their views. In theory, consultation is the opportunity for two-way communication, and can be initiated by individual citizens or their organizations as well as by the authorities. In practice, this method is afflicted with several faults such as too little information being passed on to citizens, the issues being too complex and not understandable, the time framework being too short for a thoughtful response, etc. (Johnson 1984). The following is a description of some of the most widely used administrative techniques in both the tourism and non-tourism area.

Information exchange

Information exchange is the most important step in legitimizing public participation in tourism development. If participation is to be genuine, information must be exchanged early enough and in lay terms that will allow all participants time to respond. Participation programme designers have little control over who participates, how many citizens participate, or what type of information is produced. There are basically five types of information exchange:

1 Drop-in centres – Drop-in centres can be in the form of a geographical location, such as an office within a building or a kiosk where information on the proposed tourism development plan can be collected or given to any interested party. Drop-in centres can also take the form of computer bulletin boards. In the case of bulletin boards the participants are limited only to those who have access and know-how to communicate through computers. Accessibility is a key issue with drop-in centres. Accessibility requires that the information be easily understood in lay terms and that the information is convenient to access. It is recommended that the centre be staffed by unbiased individuals who are up to date on the relevant

issues, can effectively operate a database and can identify the various participants for each issue.

2 Public hearings – Public hearings are the most common form of public participation in most western countries. Compared with other techniques, public hearings are inexpensive and easy to manage. Hearings offer all citizens, and especially those who feel that they would be affected by a proposed tourism development project, a chance to express their views and enable the decision makers to understand their concerns better. This technique, however, is not very effective for identifying alternative solutions to problems because the information flow is normally unidirectional.

3 Large and small group public meetings – These are less formal than public hearings. Unfortunately, meetings like hearings often become vehicles for one-way communication by providing superficial and/or condescending information, discouraging questions or dispensing irrelevant answers.

4 Focus group interviews – Once all parties that might be affected by a tourism development project have been identified, one or more representatives are selected from each group. The participants are interviewed as part of a round-table discussion. The interview questions are designed to stimulate conversation amongst the group members. All discussion and opinions are recorded. One problem with this technique is its lack of structure. The technique can provide an extensive list of alternatives and attributes, but does not allow for the establishment of priorities. Its major weakness is that some of the participants may dominate the discussion and influence the others with their outspokenness.

5 Telecommunications techniques – These techniques use television, radio, newspapers, e-mail, Internet and fax to send and gather relevant information. These telecommunication devices must of course be accessible to participants. Also required is participants' ability to understand the information, which in the case of tourism development might be rather technical, as well as to abstract it from all the other noise that comes through these telecommunications devices. As the term NIMBY (Not In My Back Yard) suggests, the typical citizen will participate only in those cases where they feel that a particular tourism project would personally and directly affect their livelihood, health or quality of life. Therefore the announcements or communications need to explain why people should get involved and how a certain project may affect their life.

Education and support building

Citizens may not understand tourism planning issues or realize how a certain development project may affect them. The techniques falling into this category attempt to educate the citizens and explain the relevant issues. For this reason these techniques are the second most important step in the participation process. Leaders of citizen coalitions or special interest groups may also use techniques of this type to rally support for a cause or form a power base. A major drawback of these techniques is that they do not offer any assurances that the participants' concerns will be taken into account. The following are some of the most commonly used education and support building techniques.

1 Advisory groups and task forces – An advisory group is chosen under specific or arbitrary criteria to study a specific issue, such as the introduction of casino gambling in a tourist community, and explain it to the participants. Members of the group may be representatives of specific interest groups such as, in the case of casino gambling, gaming industry officials, religious leaders, etc. These individuals have no decision-making authority, their only purpose is education.

2 Technical and professional advice – Advisers such as architects, environmental engineers, sociologists, economists and other academic tourism experts may be accountable to certain ethics as determined by their professional associations, but they are also accountable to those who hired them.

3 Petitions – These are intended to educate the powerholders about citizen concerns. The number of signatures on a petition may indicate the magnitude of the problem and the level of concern. Petitions do not reveal the whole range of alternatives, nor do they identify all those who may be affected by the issue.

4 Workshops and seminars – These are meetings where people intensely study and/or discuss various issues related to a tourism development project. Workshops may include break-out sessions which allow groups with special interests to study or discuss more specific topics – i.e. the effect of an airport location on property value. Workshops can help those who wish to become more involved in the participatory process and align themselves with specific interest groups. Workshops and seminars can study alternatives in greater detail, which in the final analysis may result in a wider consensus among all groups involved.

5 Expert panelling – In this technique citizens are invited to a meeting where tourism experts discuss the issues among themselves. These experts may be speaking in professional jargon which is foreign to some of the citizens. For this to be an effective technique the communication between experts needs to be understandable by all those who are observing the discussion.

6 Formal and professional training – Citizens enrol in a college- or university-level programme where they can learn how to develop a power base, design a participation programme, implement the whole array of participation techniques, etc. Developing a successful programme of citizen

participation in tourism planning depends on several factors, such as: the issues, the characteristics of the participants, timing, skills of the programme implementors, previous participation programmes in that locality, etc. Each planning issue is unique, therefore the participation programme that accompanies it must also be unique. Because of the complexity and the uniqueness of planning issues, participation programme effectiveness can be greatly enhanced when the designer and participants of a programme have an academic foundation of knowledge in that area.

Citizen oriented techniques

These techniques are designed to give citizens the power to make tourism planning decisions or influence the decision-making process. They sensitize the powerholders to the values of the citizens and force politicians and government bureaucrats to be more responsive and accountable to the citizens. These techniques consist of *decision-making supplements* and *representational input*. Representational input is further divided into two categories to distinguish between active and passive representation.

Decision-making supplements

While in most countries citizens do not have the authority to make decisions regarding tourism development projects, they can exert a strong influence on the outcome of decisions. Three techniques are delineated:

1 Direct confrontation – The citizens directly confront the powerholders. This may be done by verbally challenging the powerholders or by staging a demonstration against the planned development of a particular tourism project. This activity may continue until the participants get a desired response. Terrorism and sabotage are extreme forms of a confrontational technique.

2 Litigation – A citizen files a legal suit against another person or an organization. For example, in Central Florida one citizen sued the Orange County Commission for its imposition of an additional 1 per cent resort tax to be used for financing the building of a baseball stadium. This form of participation is antagonistic. It is a reactive rather than a proactive form of participation. Litigation is either resolved out of court through negotiations between constituents or by judge or jury in court. It can be very effective in blocking or changing a course of action.

3 Role playing and game playing – This technique is used to achieve consensus among citizen groups who represent different sets of values. Representatives from various interest groups such as hoteliers, homeowners, developers, etc. 'exchange shoes', so to speak, and argue the issue

169

from an opponent's perspective. This game playing is designed to enhance communication and understanding between these groups.

Representational input (active process)

This is the most potent form of citizen participation.

> It is the actual exercise of decision-making power by some duly recognized body of non-official, either by itself or shared by a local council. This type of participation is definitive in its legal authority to impose requirements, grant permission, and commit resources. Using initiative or referendum procedures, citizens can make planning decisions by popular vote. Participative relationships will be definitive to the extent that: (1) authorities make a commitment to heed citizen choices and either set formal criteria for considering them or implement them without further consideration, (2) there is mutual trust and a common political interest between authorities and participants, (3) the participants make up an established, continuing body to which power can be responsibly delegated, and (4) that citizen body is supplied with the necessary information and support to make competent decisions. The empowered citizen participants need to be accountable to its constituents to a degree that matches the authority it has.
>
> (Johnson 1984)

These participation techniques are considered to be active because the citizens are actively partaking in the decision-making process. The following are descriptions of the four most commonly used active-representational techniques:

1 Votes, referendums and plebiscites – these are the submissions of plans or laws to the direct vote of the general public. Many communities throughout the world allow citizens to initiate a referendum on any development issue including tourism, by securing signatures from a specified percentage of registered voters.
2 Partnership – A proportion of public councils/commission seats may be allocated to citizen representatives. In many tourist communities where there is a local tourist council/commission which is mostly composed of tourism industry representatives and elected public officials, it is common to reserve a few seats for citizen representatives. Therefore it is possible to say that in such cases decision making is shared by officials and citizen groups. One problem associated with this method is trying to determine who the citizen representatives will be.
3 Delegated power – Since many comprehensive tourism development plans are mostly concerned with general issues, it is possible to leave the

finer issues for citizen groups to decide. Delegated power gives the citizen body authority over particular plans or programmes.

4 Citizen control – Citizens may have complete control when local development policies are loosely stated. Some governments publish a list of recommended guidelines. But these guidelines are not enforceable and give complete responsibility for a specific area or function to the citizens.

Representational input (passive process)

Participants are representative of neighbourhoods, coalitions, public agencies or other special interest groups that have a stake in tourism development issues. Participants take part in the establishment of alternatives and the setting of priorities. This is classified as a passive process because the representatives are not responsible for making the decision. These techniques are best used for developing a consensus among the various interest groups. Before this process can be effective, the best possible effort needs to be made by the planning authorities to identify all the possible sets of actors in the tourism development issue. The representatives should be elected by members of each set of actors.

1 Nominal group technique (NGT) – It is normally implemented in six stages. The session moderator presents the participants with a statement on the topic area to be discussed. Once everyone clearly understands the topic, each participant is asked to reflect individually on the topic, and to record their personal response on a worksheet. The moderator randomly chooses participants to state their responses. This continues until all participants have responded. All ideas are consolidated. From the consolidated list of ideas, the participants rank each of the responses by level of importance. The moderator then compiles the results. The nominal group technique is a systematic approach designed to provide two specific types of output. First, it provides a list of ideas relevant to the topic question. Second, the technique provides quantified individual and aggregate measures of the relative desirability of the ideas raised in the session. The NGT requires that all participants meet at a central location. The NGT, however, can be conducted nowadays over the Internet or through a host computer. This would of course require that all participants have access to this type of technology.

2 Delphi process – Participants are asked to submit issues to the process leader. The process leader distributes the compilation of views and the participants are asked to rank them in order of significance. This method, in contrast to the NGT, does not require that participants meet in one place. The structure of this method allows for statistical analysis. This technique lends itself to phased research and planning programmes in which data are gathered sequentially.

171

3 Citizen surveys – A survey on the relevant tourism issues is constructed by a skilled researcher. The survey is distributed to previously determined participants who complete it and return it to the researcher for analysis. The survey may also be conducted in an interview format. It produces a structured output that facilitates analysis and makes use of probability sampling. Despite methodical preliminary efforts, most researchers have had difficulty acquiring sufficient data in some areas of the research problem.

4 Planning charrettes – To develop a range of planning and design ideas for enhancing a tourist destination, some cities conduct a multi-day, high-energy, exciting event which involves the public at large. The designing involves a large number of citizens working in groups with architects. Residents, tourism industry executives, merchants, architects and landscape planners work together to create a plan that architects refine and translate into drawings. The plan agreed upon is submitted to the proper authority for approval. One example of such an event is the redesign of Park Avenue in Winter Park, Florida, a major tourism destination in town. The town planners believed that inviting the public to help was the best way to create the plan because 'who better to do this than the people who work and live on the avenue, who think about it all the time and have an intimate understanding of it?' (*Orlando Sentinel* 1996: K-4). According to the planners of the Winter Park charrette, this is an exciting way to involve the public in what normally is a behind-closed-doors, drawn-out process; and a way to tap into a broad range of expertise without spending a lot of money.

It is important to note that each community is unique in its fiscal stability, leadership, tradition, legal mandates, etc. For this reason it is difficult to develop a universal set of fundamentals for citizen participation in the tourism planning process. And so effective participation programmes in tourism development projects require a combination of techniques that will work best for its unique set of constituents.

ENCOURAGING CITIZEN PARTICIPATION IN TOURISM DEVELOPMENT

Encouraging citizens to participate in tourism development will always be a difficult task. This requires, first and foremost, the opening of power distribution channels and legitimizing issues. There are two types of barriers that affect power distribution. One type is associated with the powerholders, who will resist distributing this power unless they feel they will get something in return. The other type is associated with the have-nots. The barriers include, first, inadequacies in the citizens' political and socioeconomic infrastructure and knowledge base; and second, difficulty in organizing a representative and accountable citizens' group in the face of the futility, alienation and distrust that many citizens feel nowadays in relation to their governments (Arnstein 1969: 217).

Information tends to be restricted in the early stages of planning. Tourism entrepreneurs, like all other business people, exercise secrecy to protect their plans from competitors. The closed bargaining sessions between executives and local officials in the early part of tourism planning prevents citizens from participating early enough to have any influence. By restricting the flow of information, citizen participation becomes merely 'tokenism'. A lot of money and time is invested in preliminary research before a business even approaches local officials with its proposals. Policies which would open up the bargaining tables to citizens in the early planning stages might scare off potential investors. Communities that are hungry for growth impose few restrictions on investors. Later, when growth nears saturation point, officials may become more strict about what developments take place.

Developing a policy which is part of a comprehensive plan that allows citizen participation at an early stage without scaring off potential investors is an area for future research. One possible solution to this problem is to establish an incentive programme for investors/developers. To receive certain incentives these investors/developers must adhere to specific participation programme requirements. In the case where the money is coming from a federal or state level, federal or state institutions can mandate a policy that would require community officials to submit all planning proposals, including those for tourism development projects, for citizen approval. If a proposal is vetoed, it could delay funding until an alternative plan is approved by the citizens.

Mentioned earlier was a lack of a sufficient technical knowledge base among the citizens which interferes with citizen empowerment in the participation process. By appointing a person in the local government whose sole responsibility is to make information regarding planning decisions more easily accessible to citizens, this hurdle may be overcome. The information would include press releases, brochures and a referral service for consultants and facilitators. Citizens must perceive involvement as convenient, or they will become discouraged when their perceived inconvenience exceeds their desire to participate. So improving accessibility is the key to encouraging participation.

There are situations where citizens are involved in the tourism planning process but become discouraged when they feel that the information available to them is not thorough or is untruthful. One possible solution to this problem is the establishment of a fund that would allow citizens to hire their own consultants and facilitators who would be accountable to them only. This fund could be supported by application fees and/or impact fees. Another problem with power distribution occurs when large tourism operations, backed by deep pockets and talented negotiators, monopolize a community's tourism industry. Enforcing planning laws that restrict the size of the tourism enterprise is one way to overcome this problem. In some rural Costa Rican communities the number of rooms per hotel is restricted. Restricting the scale of an operation puts local operators on even terms with outside operations. Giving residents the opportunity to compete for tourist dollars legitimizes the tourism industry, thus encouraging participation.

LEGITIMIZING THE ISSUES

Many citizens will refuse to participate in tourism planning unless it affects their personal income, health or safety. Although tourism has both direct and indirect benefits, the indirect benefits go largely unnoticed. This does not mean that citizens do not enjoy the benefits, it merely means they take them for granted. If citizens do their part to develop sustainable tourism, that tourism must embody those values that are most important to them.

The first step in legitimizing the tourism industry in a community is incorporating tourism planning with community planning. Second, citizen participation must be written into the policies which guide tourism/community development. By embedding citizen participation in the planning policies of the community, public officials become accountable to the citizens when citizens are not adequately involved. Without a documented policy for citizen participation, citizens have no ground for protest.

In Fiji, citizens staged a protest against a resort owner's misuse of the community's resources. Effluent from the resort was destroying the lagoon, and the resort was consuming a large portion of the potable water. The citizens did not perceive the allocation of costs and benefits of tourism to be fair. Realizing that business would be hurt by the unfriendliness of the islanders, the resort owners agreed to share a percentage of the revenue with them. The money could be spent by the islanders as they wished. Getting income from the tourism industry legitimized their need to participate. It also gave them the financial resources to improve the quality of their community. Funds for such incentives can be raised through taxation, lease agreements, impact fees or stock dividends (Lea 1996).

When participation programmes are hit-and-miss, citizens can become detached from the process of ongoing planning. Efforts to keep citizens updated on various tourism developments can generate a momentum among participants for continued involvement. The person with responsibility for developing participation programmes cannot get bogged down with other duties; they must be dedicated to that task alone. There is a special need to involve not only the politically active citizens but also those who are unorganized, non-joiners or inarticulate. Without their input, a participation programme designer cannot know what modifications are needed to make the programme more effective. The same is true for the business people or local authorities that need to know whether they have considered all the alternatives and discovered all possible impacts. Unfortunately no-one knows exactly why non-joiners do not participate. One way to legitimize non-joiners' involvement in the tourism development process is by paying them. A fund can be established through impact or application fees. Paying people a nominal fee to participate encourages a fairer representation of citizen participants. While such payments might raise ethical and legal problems and establish new norms for paying citizens to exercise their duties and obligations, it might be appropriate to point out that such payments already exist in some communities. In many countries where the legal practice requires jury service, those who serve as jurors are paid nominal fees to compensate for time lost away from work. The same is true for army reserve and other emergency duties.

Participation programme evaluation

A good tourism participation programme will include evaluation of the programme's effectiveness. This provides feedback to the programme designer. How good this feedback is depends on the framework in which the evaluation is done. The evaluation should examine the context in which participation occurs, the quality of each participation procedure, the outcome of each involvement and the effectiveness of the techniques in combination. The following is a format for the evaluation of public participation that has been suggested by Smith (1984: 257).

Context:

1   Historical background
2   Institutional arrangements:
        political structure and processes
        legislation and political regulations
        administrative structures
3   Agency features:
        status
        function
        terms of reference
        financial arrangements

Process:

1   Goals and objectives for participation:
        mandate given participation by the agency
        objectives of participants
2   Number and nature of public involved:
        who are they?
        how representative are they?

how organized are they?
3   Methodology employed:
     techniques
     information access
     resources

Outcome:

1   Results of participatory exercise
2   Effectiveness:
     focus on issues
     representativeness of participants
     appropriateness of participants
     degree of satisfaction expressed by participants
     degree of awareness achieved
     impact and influence of participation
     time and cost

This approach is biased toward how well the participation programme accomplished the agency's objectives. Because of the multi-goal nature of citizen participation, the success of the programme cannot be judged from one agency's perspective alone. In a participation programme each constituent may have their own objectives for participation. A higher level of confidence can be achieved when the programme's evaluation is composed of the sum of all participants' independent evaluations. Evaluation of this nature can be done through exit interviews. During this interview participants are asked to express their degree of satisfaction. This input is then compared to a participant's objectives as personally stated at the onset of the participation technique.

Programme designers need to ensure that citizens develop realistic expectations for participation in tourism development projects. The disappointment citizens may feel because of unrealistic expectations can be harmful to the credibility of the participation programme. Participants may be inexperienced in presenting their views. Patience and understanding on the part of the programme coordinator can help extract valuable concerns and information from those who are not proficient communicators. The coordinator must be sensitive enough to see when participants are becoming overwhelmed with technical jargon. The coordinator must also make sure that the public clearly understands the issues and possible alternatives. Follow-through with the participation programme is also important to winning citizens' trust. People like to see the impact of their contribution. An important part of follow-through is keeping past participants in tune with the plan's progress.

CONCLUSION

The overview of participatory techniques included in this chapter is designed to assist tourism planners and decision makers in the preparation of a tourism participation programme. The techniques have been categorized to establish a framework for comparison. Overcoming the obstacles of citizen participation is not easy. Each community must find its own way when involving citizens in the planning process. Citizen participation can yield substantial rewards. These include an opportunity to improve the management of the community's tourism life cycle, an improved understanding of the relevant elements in the community having an impact on tourism, better anticipation of the internal and external challenges to tourism, a chance to ameliorate detrimental impacts, and a superior opportunity to accommodate the full range of publics that may be affected by tourism (Haywood 1988).

In order for tourism to be legitimized, it must be incorporated into the community planning structure. Measures need to be taken to encourage direct citizen involvement in tourism planning. Evaluation of participation programmes is critical to their survival. Many failures of citizen participation programmes in western democracies can be directly attributed to the absence of an adequate evaluative framework. As the pressure for tourism growth gathers strength in many parts of the world, the need to promote sustainable tourism development through citizen participation becomes an important element in the planning process.

REFERENCES

Arnstein, S. R. (1969) 'A ladder of citizen participation', *AIP Journal*, 35: 216–24.

Brougham, J. E. and Butler, R. W. (1981) 'A segmentation analysis of resident attitudes to the social impact of tourism', *Annals of Tourism Research*, 8 (4): 569–90.

Burke, E. (1979) *A Participatory Approach to Urban Planning*, New York: Human Sciences Press.

Burton, T. L. (1979) 'A review and analysis of Canadian case studies in public participation', *Plan Canada*, 19 (1): 13–22.

D'Amore, L. J. and Murphy, P. (1983) 'Guidelines to planning in harmony with the host community', in L. J. D'Amore and P. Murphy (eds) *Tourism in Canada: Selected Issues and Options*, Western Geographical Series, vol. 21, Victoria BC: University of Victoria.

Getz, D. (1986) 'Models in tourism planning towards integration of theory and practice', *Tourism Management*, 7 (1): 21–32.

Glass, J. (1979) 'Citizen participation in planning: the relationships between objectives and techniques', *APA Journal*, 45: 180–9.

Gunn, C. A. (1993) *Tourism Planning*, 3rd edn., Washington DC: Taylor and Francis.

Hall, C. M. (1994) *Tourism and Politics: Policy, Power, and Place*, London: Belhaven Press.

Haywood, M. K. (1988) 'Responsible and responsive tourism planning in the community', *Tourism Management*, 9 (2): 105–18.

Johnson, W. C. (1984) 'Citizen participation in local planning in the UK and USA: a comparative study', *Progress in Planning*, 21 (3): 149–221.

Lawson, F. and Baud-Bovy, M. (1977) *Tourism and Recreation Development: A Handbook of Physical Planning*, London: The Architectural Press.

Lea, J. P. (1996) 'Tourism, realpolitik and development in the South Pacific', in A. Pizam and Y. Mansfeld (eds) *Tourism Crime and International Security Issues*, London: Wiley, 123–42.

Long, P. and Nuckolls, J. (1993) 'Organizing resources for rural tourism planning, and technical assistance', *Tourism Recreation Research*.

Loukissas, P. (1983) 'Public participation in community tourism planning: a gaming simulation approach', *Journal of Travel Research*, 22 (1): 18–23.

Marsh, N. R. and Henshall, B. D. (1987) 'Planning better tourism: the strategic importance of tourist-resident expectations and interactions', *Tourism Recreation Research*, 12 (2): 47–54.

McIntyre, G. (1993) *Sustainable Tourism Development: Guide for Local Planners*, 1st edn., Madrid: WTO.

Murphy, P. (1981) 'Community attitudes to tourism: a comparative analysis', *Tourism Management*, 2(3): 189–95.

——(1983) 'Tourism as a community industry: an ecological model of tourism development', *Tourism Management*, 4: 180–93.

——(1985) *Tourism: A Community Approach*, New York: Methuen.

——(1988) 'Community driven tourism planning', *Tourism Management*, 9 (2): 96–104.

National Economic Development Council's Tourism and Leisure Industries Sector Group (1991) *The Planning System and Large-Scale Tourism and Leisure Developments with Case Studies*, London: NEDO.

Orlando Sentinel (1996) *Residents Will Help Create Park Overhaul*, 24 March, K, 1–4.

Rees, W. E. (1989) 'Defining sustainable development', *CHSR Research Bulletin*, University of British Columbia, May.

Ritchie, J. R. B. (1985) 'The nominal group technique: an approach to consensus policy formulation in tourism', *Tourism Management*, 6 (2): 82–94; 9 (3): 199–212.

Sewell, W. R. D. and Coppock, J. T. (1977) *Public Participation in Planning*, London: Wiley.

Sewell, W. R. D. and Phillips, S. D. (1979) 'Models for the evaluation of public participation programmes', *Natural Resources Journal*, 19, 337–58.

Smith, L. G. (1984) 'Public participation in policy making: the state-of-the-art in Canada', *Geoforum*, 15 (2): 253–9.

So, F. S. and Getzels, J. (eds) (1988) *The Practice of Local Government Planning*, 2nd ed., Washington DC: International City Management Council.

Warren, R. L. (1969) 'Model cities first round: politics, planning, and participation', *AIP Journal*, 35 (4): 245–52.

Williams, P. W. and Gill, A. (1991) *Carrying Capacity Management in Tourism Settings: A Tourism Growth Management Process*, Alberta: Centre for Tourism Policy and Research, Simon Fraser University.

Williams, T. (1979) 'Impact of domestic tourism on host population: the evolution of a model', *Tourism Recreation Research*, 4: 15–21.

Wolf, J. (1979) 'Evaluations of objectives: a case study and commentary on public participation', *Plan Canada*, 19 (1): 38–48.

World Commission on Environment and Development (1987) *Our Common Future*, New York: Oxford University Press.

11

IMAGE AND SUSTAINABLE TOURISM SYSTEMS

William C. Gartner

INTRODUCTION

The formation of touristic images and their relationship to sustainable tourism systems (STSs) are not easily understood or directly observable. The reason for this has more to do with the ambiguous meaning of the term/concept sustainable than anything else. Elsewhere in this book other contributors have addressed the meaning of sustainability and STSs, but to understand how the concept is related to tourism image it is necessary briefly to review some of the controversy surrounding what is meant by sustainable.

Sustainable is a value positive word. Simply put, it has such strong, positive connotations that it is easily and eagerly embraced by almost all members of society. Its root meaning is to nourish or prolong. Problems arise however when people begin to use the concept in an operational context. For example, assume a resort destination area has been experiencing double-digit growth rates for the last five years. For developers, sustainable could easily mean maintaining (i.e. prolonging) those growth rates into the foreseeable future. For members of the host community the concept may mean growth must be significantly reduced to maintain (nourish) the quality of life, including off-season periods of recuperation, that the residents currently enjoy. Both groups believe in the concept of sustainable tourism systems but disagree on how it is interpreted and applied. Even the most widely accepted definition of sustainable, which involves continued use without depletion of resources (environmental or human) is problematic. Many argue that change is a constant. Even without human intervention natural systems change as they evolve. Therefore, embracing a concept that disallows change is contrary to natural law.

Sustainable development concerns grew out of the rapid exploitation of the earth's environment for economic gain. Realization that short-term gains were coming at the expense of long-term losses for future generations began to reshape global development strategy. Dragicevic (1991) traces concerns over sustainable development and the environmental consequences of

179

development back to the late 1950s. These early warnings were generally ignored until the World Conservation Strategy of 1980 (International Union for Conservation of Nature and Natural Resources 1980), the Bruntland report of 1987 (United Nations 1987) and the Charter on Sustainable Tourism arising from the World Conference on Sustainable Tourism reinforced the call for action. All recommend a global sustainable development strategy that encompasses all potentially exploitative economic development activities, including tourism.

In spite of all the attention it has received, sustainable development is still a concept searching for a clear definition. While it is risky to paraphrase the advice given by international experts, definitions proposed by Pearce and Turner (1990: 24) and action strategies such as found in Inskeep (1990) deal primarily with the basic premise that all groups involved in tourism should become more aware of development's environmental and sociocultural impacts and strive to reduce undesirable byproducts of tourism activity. Given that actions to accomplish this aim would result in less intrusive forms of tourism development, there is still no clear definition as to what sustainable development is, and how sustainable tourism systems change over time to accommodate population increases or increasing tourism demand.

What does all this have to do with tourism image? During the last twenty years image research has been an active field. Many researchers have studied how images change over time and the stimuli behind the change, and some have attempted to model the image formation process. Destination images, regardless of how they are formed, play an important and powerful role in an individual or group's travel decision process. Since sustainable development is a function of use and destination images can be used to increase use, there appears to be a direct link between how images are formed and the concept of sustainable tourism systems. As will be discussed, there is indeed a link between the two but it is far from direct and it does not follow that images used to increase visitation make achieving an operationally sustainable system more difficult. However, before delving into that line of reasoning, some background on the image formation process is necessary.

IMAGE FORMATION

We all hold images of places. When asked to provide a mental picture of a place most people are able to do so. Even if one is not familiar with a place, further information as to the country or area of the world in which it is located should elicit a response. The process of image formation then involves not only creating awareness of place but projecting selected images to an identified audience or market segment that is deemed most receptive to the message embodied in the images.

Gunn (1972) was one of the first to break the image formation process into component parts. He suggested that images were formed on two levels which he termed induced and organic. Induced image formation is a function of the marketing and promotion efforts of a destination area or business. Pictures, as well as the written material produced and disseminated by destination promoters, attempt to form images in the minds of prospective travellers. These images may be projected to people through the media, via paid advertisements or direct to individuals assumed to be predisposed to travel to the area.

Organic images emanate from sources not directly associated with any development organization. News reports, movies, newspaper articles and other ostensibly unbiased sources of information generate organic images of places. The underlying difference between an induced image and an organic one is the control that people in the destination area have over how the image is presented. Induced images are presented by destination promoters and project what the destination is offering to travellers. Organic images are not directly controlled by destination promoters but are some other entity's idea of what exists there.

Phelps (1986) contends that images are formed on two levels, primary and secondary. Secondary image formation results from any information received from another source. Primary image formation results from actual visitation. In a sense Phelps has grouped both Gunn's induced- and most of the organic image formation agents into one type and separated out actual visitation into a distinctly different form of image formation.

To understand the relationship between the image formation process and sustainable tourism systems it is necessary to analyse further the different ways in which destination images are formed. Using Gunn's image typology as a starting point, it is possible to view the image formation process as a continuum consisting of eight distinctly different components. The stages often operate concurrently, forming an image in the mind of the prospective traveller that is individually distinct. In other words, an individual's perceptual filters will form an independent image unique to that person but with features shared by others.

OVERT INDUCED I

Beginning at the induced end of the continuum, the first image formation agent is termed Overt Induced I and consists of traditional forms of advertising. The use of television, print media, brochures, billboards, etc. by destination promoters is a direct attempt to form particular images in the minds of prospective travellers. The person receiving the message is not confused by who is delivering it. It is clearly a blatant attempt to construct an image of the salient attributes of the destination in the minds of the targeted audience. Depending on the type of medium chosen, the cost of reaching an individual may be very low although total cost can be quite expensive.

Overt Induced I image formation agents have the advantage of being able to achieve widespread coverage (market penetration) as well as targeting specific markets. But they suffer from low credibility problems. People are constantly being subjected to advertisements for all types of products. Many of the touted product attributes are not always as good as the advertisements lead one to believe. It does not take too many experiences with products that fail to meet expectations before consumers become sceptical about what they are told. Tourism advertising is no exception. People may even be more dubious about some of the projected images of destinations simply because of the cost involved in trying out the product.

Most of the tourism Internet and Web technology in use today can be categorized as a form of Overt Induced I image formation. Use of home pages identifies the producer, although producing the page is much less costly than traditional forms of advertising. At the moment, however, informing a potential customer of the existence of a home page and its address is becoming more problematic, as other forms of advertising are often used to direct someone to a site on the Web.

OVERT INDUCED II

Overt Induced II image formation agents consist of information received or requested from tour operators, wholesalers and organizations which have a vested interest in the travel decision process but which are not directly associated with any particular destination. Tour operators act as gatekeepers of information, with the type of information distributed by them contributing to the images people hold about certain areas (McLellan and Noe 1983; Bitner and Booms 1982; Murphy 1983). A major function of tour operators is to create attractive destination images for the places to which they arrange tours. Destination promoters will work with tour operators to select images to be presented, but since tour operators are interested in increasing their business, only certain images will be passed on to their clientele. This may lead to unrealistic portrayals of place and result in destination images not supported or desired by a destination's host society.

Destination area promoters do have some control over the images presented through tour operators, especially in countries where operators must register. When this is the case, tour operators may be subject to pressure, subtle or direct, to project specific destination images. Most of the time there is no conflict, as national organizations are primarily interested in marketing the country and increasing visitation. Images which tour operators use to achieve this goal will most likely be supported by the national tourism organization. Because of the credibility tour operators enjoy with their clients, this source of image formation may surpass the importance of all the Overt Induced I forms, especially in countries where foreign travel is heavily dependent on package tours.

Countering the high credibility tour operators have with their clients is lower market penetration. Independent tour operators or wholesalers do not have the resources to utilize fully Overt Induced I types of image formation agents, with most concentrating on speciality markets. Therefore only selected markets will be exposed to the image formation campaign.

COVERT INDUCED I

Proceeding along the image formation continuum, the next component is termed Covert Induced I. It consists of developing destination images using traditional forms of advertising, as in Overt Induced I; however, the image is now being projected through the use of a second-party spokesperson. The use of a second party tends to mask attempts by destination promoters to influence the audience directly, hence the use of the adjective covert. Covert Induced I types of image formation are a direct attempt to eliminate some of the problems of low credibility inherent in the Overt Induced I image formation process. Second-party spokespersons are generally chosen based on name recognition and credibility. Credibility with the market is therefore improved, but the trade-off comes in terms of increasing cost since the second-party spokesperson is usually compensated for their time and audience recognition factor. The second-party spokesperson approach works best when specific products are advertised, but it can also be used for tourism image development if the second party has positive, high-recognition value.

A recognizable spokesperson will attract attention to the endorsed product using their attractive and likeable qualities to differentiate their advertisement from the clutter of other advertising messages (Atkin and Block 1983). Credibility is enhanced whenever the advertised product has high psychological or social risk (Friedman and Friedman 1979), which may be the case with long-haul international travel. The use of a second party does not affect market penetration because the same advertising strategy as that in Overt Induced I is involved.

COVERT INDUCED II

The fourth component of image formation is termed Covert Induced II. The person influenced by this agent should not be aware that destination promoters are directly involved in the development of the projected image. Covert Induced II agents take the form of ostensibly unbiased articles, reports and stories about a particular place, delivered by someone with high credibility who apparently has no vested interest in the destination. Familiarization trips are generally the vehicle used to achieve Covert Induced II types of image formation. Travel writers for newspapers, magazines or specific activity groups may be invited to participate in an all-expenses paid trip to some area to sample the attractions. The end result is that writers facing deadline pressure will often write about their most recent travel experience. This type of advertisement is a form of image development which may be very useful in reaching identified target markets. Familiarization trips are also held for tour brokers and operators in an attempt to develop favourable images in their minds which will then be passed on to their clients.

Credibility rises as the image is now projected through a source that does not appear to have any connection with the destination except through visitation. Cost is less than with other forms of induced image formation, as only expenses related to the writer's visit are covered, with production and distribution costs of the image development effort borne by the publication purchasing the writer's work. The increase in credibility and reduction in cost are somewhat offset by a decline in market penetration. For the image formation effort to be useful, the intended audience must be exposed to the report, article, etc. that has been produced. This requires that a reader be predisposed to learn more about the area or specific activity featured. There is no captive audience with this type of image formation.

The destination area has less control over the published travel account because it has no veto power over what is written. Because of this, the type of image projected may not match the type of image with which residents of the destination area wish to be presented. However, Covert Induced II types of image formation provide a relatively low-cost means to develop destination images, albeit with some risk. For example, resort areas and small communities with limited advertising and promotion funds may find familiarization trips for carefully selected participants to be an effective way to develop their tourism image. It should be noted that some newspapers and magazines refuse to allow their travel writers to participate in an expenses-paid *familiarization* trip, in order to avoid accusations of biased reporting which may also affect other aspects of their operation.

AUTONOMOUS

The fifth image component is termed Autonomous. It consists of independently produced reports, articles, films, documentaries, etc. about specific places. There are two sub-components in the Autonomous category: News and Popular Culture.

The most common form of autonomous image formation agents are television news stories. The destination area has no control over what appears in the story and its image is subject to someone else's interpretation. News stories, because of their apparently unbiased approach, are assumed to have major impacts on tourism image development. If the event reported is of major consequence, the opportunity for image change in a relatively short period of time presents itself.

For example, Gartner and Shen (1992) studied the impact of the Tiananmen Square events on the Peoples Republic of China's (PRC) tourism image. A study assessing the mature markets' image of the PRC was conducted prior to Tiananmen Square. After the incident was reported on, sometimes with live coverage, by every major news station in the free world, the study was repeated, one year after the conflict, using essentially the same sample population and an identical survey instrument.

There were twenty-two activity and attraction attributes grouped into five categories which were analysed for change. Interestingly, even though most of the attributes did show declines in impressiveness ratings, only seven out of twenty-two were significantly lower (at the 0.05 level of probability). The researchers also investigated image change for ten tourism services. Again, even though all ten declined in absolute value terms, only four were significantly different. Significant differences were noted for services that were most likely to be important to tourists, such as safety and security, receptiveness of local people to tourists and pleasant attitudes of service personnel. Survey respondents did not view services affecting the efficient movement of people (e.g. on-time arrivals and departures, inland transportation) as having changed as much but did consider people-to-people relationships to have suffered as a result of Tiananmen Square. In conclusion, the autonomous image formation agents did change the PRC's tourism image in the short run, with the nature of the change more related to how the provision of touristic services were perceived rather than the country's activity or attraction base.

Another example of what can happen as a result of major news coverage is provided by Roehl (1990). He investigated changes in travel agents' attitudes after Tiananmen Square. He found many agents advocated trade and military restrictions after the conflict. As attitudes are related to image formation, especially in the cognitive state, future travel to the PRC may depend not only on images held of touristic attitudes, but also on attitudes toward the host government.

Other studies also support the influence of autonomous image formation agents. The US Travel Service (1977) concluded from a study investigating foreigners' perceptions of the United States, that many images were based on news reports and movies that depicted violence in the country. The *Pacific Travel News* (1984) reported that US citizens' images of Korea were primarily derived from the popular television series *M.A.S.H.* and represented an image of Korea dated more than thirty years ago.

The autonomous image formation agent, because of its high credibility and high market penetration, may be the only agent capable of changing an area's image dramatically in a short period of time. One of the reasons for this may be the lack of information people have about destinations that are far removed from their home residence. If people form images based on little acquired information they are more susceptible to change when massive amounts of information are processed in a relatively short period of time.

The above studies provide evidence supporting the importance of news reporting as autonomous image formation agents. It is also possible that the effect of negative autonomous change agents, although significant in the short term, may not be an important factor in long-term image change. Thurstone (1967) contends that, in the absence of any reinforcing information, images may revert to those held before the exogenous shock. Although not thoroughly studied, this does appear to be the case, as in the absence of recurring events (e.g. Tiananmen Square) visitation figures show a rebound after the event has become part of history. The amount of time it takes to rebound does show some relationship to the event's magnitude and its cause. Natural disasters have short-term effects rarely lasting more than one season. Events instigated by people show longer periods of recovery. What does seem clear is that in the absence of reinforcing negative images there is a reversion to previously held images.

Popular culture portrays images of people and places. Increased travel to certain countries has been correlated with the success of films that use country images as a major backdrop to the story's plot. Although the effects of popular culture on image formation have not been thoroughly studied, there is enough anecdotal evidence to suggest a strong relationship exists. Many states in the US have even organized film promotion offices to encourage companies to select locations in the state for their movies, television shows, etc.

UNSOLICITED ORGANIC

The sixth image component is termed Unsolicited Organic. It consists of unrequested information received from individuals who have visited an area or believe they know what exists there. Information on other areas is received by people on a regular basis in conversation over coffee, at dinner with friends, during business meetings or in any place or setting where the topic shifts to world politics or just simply places recently visited. The person receiving the information has not requested it and therefore the credibility factor is only moderate. However, since it comes from an acquaintance it may carry a higher level of credibility than information received from any of the induced agents. Low market penetration counters the higher level of credibility, since only those people exposed to the message will be affected as individual communication is much less pervasive than mass media. But if the information is about a destination that has received relatively little exposure through the induced or autonomous agents, then unsolicited organic information can be a very important source of image formation. Cost of unsolicited organic image formation is nonexistent unless one considers the indirect cost to the destination of an unfavourable report.

SOLICITED ORGANIC

The seventh component is termed Solicited Organic. It consists of requested information received from a knowledgeable source, generally one's friends or relatives. Because of the nature of the information flow, i.e. someone responding to a specific information request, the credibility factor is very high. Solicited information is often used to move a destination from the awareness set into the inert, inept or evoked set of the travel decision process (see Howard 1963; Campbell 1969; Narayana and Markin 1975; Woodside and Sherrell 1977; Thompson and Cooper 1979, Woodside and Ronkainen 1980, for a more complete review of the travel decision process and different sets). It may also be used to help select a destination from the evoked set. The solicited organic component is very important when a person is in an information search mode. This stage of image formation is also referred to as 'word of mouth' advertising. Market penetration is very low, however, because as more and more people are contacted for information a point is reached where most people will soon be providing information that simply reinforces what was received from others. Stutman and Newell (1984) refer to this as acquiring salient beliefs. When the point where substantial reinforcement of previously stated beliefs is reached, the information seekers will analyse the information, evaluate it in terms of their own beliefs, assess motivations and finally make a decision as to which set to move the destination into. Although there is no direct cost to the destination area for this stage of image formation, the experience provided by the destination is of critical importance in forming positive salient beliefs about travel to the area. In the long run, in the absence of any negative autonomous image formation, the solicited organic component is the most critical determinant of an area's economic tourism health.

ORGANIC

The final component of image formation is simply termed Organic and consists of actual visitation, after which a new destination image is formed in the mind of the visitor. The visitor, holding a new image, feeds back into the image formation cycle as a distributor of information in the unsolicited and solicited organic components. Many studies of image change as a result of visitation have been conducted (Goodrich 1978; Phelps 1986; Gyte 1988; Shen 1989; Gartner 1989; Khan 1991, among others).

DESTINATION IMAGE CHARACTERISTICS

Having examined the different ways in which destination images are formed, there are some general image change characteristics that more or less guide those interested in developing and manipulating images. They are:

Touristic images change slowly with the larger the entity the more slowly the image changes. The rate of image change is inversely related to the complexity of the system. This is as true of complex ecosystems as it is of socially and politically structured entities (e.g. nations, cities, communities). A tourism image is made up of many different parts including the natural resource base on which activities take place, the sociocultural system that governs the provisions and type of touristic services, and the created structures that serve the needs of tourists and may also provide some of the attractions. Boulding (1956) suggests that information affecting held images can produce three effects.

In the first instance, information is received that is not in conformity with held beliefs, setting up a situation of cognitive dissonance. The individual will attempt to avoid the incoming information, thereby reducing the dissonance. If enough information can be avoided the image remains essentially unaffected. In the second instance, the information keeps coming and cannot be avoided, resulting in a gradual image change. In the third instance, enough new information is received to result in a general reassessment of the image previously held, and leads to an entirely new image. So the key element in image change is the amount and extent of new information which is in contrast to the image currently held. Autonomous image change agents, if constant and prolonged, will eventually be unavoidable, causing an image shift. Induced image formation agents can also have the same effect, but because of their low credibility rating will take longer to affect change.

Gartner and Hunt (1987) studied touristic image change for the State of Utah over a twelve-year period. There was evidence linking increased visitation to an improved state image, although Utah's image did not improve dramatically with respect to its nearest competitors. The image of the region in which Utah was located had also improved, leading to the conclusion that an improvement in a larger entity, the inter-mountain region of the United States, had benefited all states within the region.

Further evidence of the relatively long time factor involved in changing touristic images can be found in Crompton (1979), Cumings (1983) and Kotler (1982). In the absence of any major news event causing an individual to process massive amounts of information quickly, destination image will remain relatively constant in the short run.

Induced image formation attempts must be focused and long-term.
As a result of the time it takes to change an image, any induced image
formation programmes must be focused and long-term. If destination
promoters have scarce financial resources, that fluctuate on an annual basis,
and the image change effort is on again/off again, they would be better off
focusing on improving their product and utilizing the organic image formation agents for promotion. Images tend to have stability (Crompton and
Lamb 1986) and as discussed above, in the absence of any major autonomous
impacts, will take years to shift. Therefore consistency is a requirement for
long-term image change using induced formation agents.

Consistency should not be confused with repetition of the same message.
Exposure to the same advertisement can result in a diminishing marginal
rate of effectiveness (Schumann *et al.* 1990). Six exposures to the same
commercial may be the point at which diminishing returns occur (Grass and
Wallace 1969). Consistency in this case refers to the planned long-term
delivery of a message that has at its core a common theme (image) without
using the same vehicle to deliver the message.

**The smaller the entity is in relation to the whole, the less of a chance
to develop an independent image.** This rule has its exceptions, and
generally they relate to distance from market and strength of brand image.
Hunt (1971) in one of the earliest tourism image studies, found that
distance from one's permanent residence to destination was a factor in the
image held of the destination. The further one lives from the destination in
question, the less likely one is to have a clear image of the destination.
Crompton's (1979) study of images of Mexico lends further support to
Hunt's earlier conclusions.

Khan (1991) studying the image of Wisconsin's tourism regions and
comparing those images to the prevailing state image, found with very few
exceptions that regional image matched state image. The subjects in Khan's
study were non-residents of states surrounding Wisconsin. These findings
lend further support to the strength of brand image, in this case a state
image, overpowering images of smaller entities located within the state.

A brand, as defined by Okoroafo (1989) is a name, design, symbol or
combination of these used to identify a service or product. A state, or in the
case of international tourism, a country, provides brand identification for
image development. Slogans such as 'I Love New York', 'Akwaaba' (you are
welcome – Ghana) or 'Say Yes to Michigan' all create a point of reference and
create a perception based on the images evoked from these brand identification statements. Brand identification does not rule out the establishment of
a strong independent image at the community or regional level, but the
establishment of that image will be more important in localized areas and
less distinct the further one is removed from the community. Communities

or regions can use this to their advantage, however, as they may be able to piggyback on a strong brand image in their advertising and promotion.

As mentioned, there are exceptions, and it is possible for a smaller entity to have such a strong image that it overpowers the larger area. This may be the case where the only opportunity for developing an image is through the autonomous agents. For example, if the results of the US Travel Service study, which found foreigners' images of the United States were a function of news and movies depicting violence in the country, were extended, it would not be unlikely to find that images of certain states were related more to images held of its cities. In this case cities receiving the exposure are considered the brand against which the other areas in the state are evaluated.

Image change, to be effective, depends on an assessment of present images. Changing an image depends on knowing what images prospective travellers now hold and initiating efforts to reinforce existing images or move images in a new direction. Image change efforts are essentially wasted if baseline data establishing present image position are not known. Understanding images held by target markets is essential to avoid moving the image into a position held by an able and strong competitor.

(For a more complete review of the image formation process see Gartner 1993, 1996)

IMAGE AND SUSTAINABLE TOURISM SYSTEMS

To achieve a sustainable tourism system, host communities must become directly involved in identifying the appropriate images to project, and must decide which groups will be the recipients of the images. They must also decide what they mean by sustainable, and then determine if their resources are sufficient to support the images they wish to project. The reason for this is that images build experience expectations and different touristic experiences will tax local systems in different ways. Cohen (1978) discusses four fundamental factors that affect environmental impacts. They are resiliency of the ecosystem where the activity takes place, intensity of site use, type of activities engaged in, and length of stay. In the absence of any proactive planning, the chances for achieving a sustainable system diminish.

One can look at the process as a series of steps that must be addressed concurrently. Basically, the process begins by answering questions which will provide a good idea of the current situation and the desired outcome. Some of the questions that should be asked include:

1 What type of tourist do we now attract?
2 Is this the type of tourist we want to attract in the future?
3 What is the image we currently project?
4 How do we project our current image?

5 What type of tourism development planning is currently in place?
6 Are there any indications that the type of tourism we have is over-whelming any of our local systems (e.g. environmental)
7 What indicators can we use to measure change?
8 Who is responsible for measuring change?

Once these questions are addressed then an appropriate image mix can be determined. The information contained in Table 11.1 can be used to help select the right mix. Particular attention should be paid to the columns on market penetration. The lower the market penetration generally, the slower images will change. If this is the type of image formation currently employed, the destination has the best opportunity to adapt to change as it should occur slowly enough for corrective measures to be taken when necessary. The organic end of the continuum is associated with low levels of market penetration, and for most small communities these may be the best types of image formation agents to use.

However, if a community is tourism-dependent or the image it currently projects is not the one it wishes to project, the right image mix may contain some of the agents at the induced end of the continuum. Extensive use of these agents should be viewed with caution. Because many of them have the ability to reach large numbers of people, the chance is present of increasing use beyond a destination's ability to cope with growth. To achieve anything resembling a sustainable system, intensive planning, including relevant research to assess and monitor systems, is required.

Destinations that rely heavily on the induced image formation agents enter into a high-risk game with respect to sustainable tourism systems. As an analogy, consider theme parks or amusement parks. Those that prosper are always introducing new attractions at periodic intervals. Whether it is a new type of roller-coaster, wide-screen cinema, aquarium or some other type of mega-attraction, each one has its own lifespan with respect to increasing attendance from existing patrons or attracting new clientele. At some point a new attraction is required to provide an attendance boost. The same can be said for destinations heavily dependent on tourism for their economic well-being. Those destinations that rely less on natural attractions and more on contrived or human-built facilities will be more dependent on induced image formation agents to maintain a healthy tourism industry.

Little has been said about the impact of autonomous types of image formation. Even though destination areas have little control over these agents, there still exist opportunities to make use of them. For example, in the case of a natural disaster the resulting news coverage will portray an area in distress, with one of the outcomes an inevitable decline in visitation, at least in the short term. 'Sustainable' then takes on an entirely new meaning, and for some marginal businesses needed for product diversity it becomes synonymous with survival. There are ways to manipulate the media and

Table 11.1 Image formation agents

Image Change Agent	Credibility	Market Penetration	Destination Cost
Overt Induced I Traditional forms of advertising, (e.g. brochures, TV, radio, print, billboard, etc.)	Low/Medium	High	High
Overt Induced II Information received from tour operators, wholesalers	Medium	Medium	Indirect
Covert Induced I Second party endorsement of products via traditional forms of advertising	Low/Medium	High	High
Covert Induced II Second party endorsement through apparently unbiased reports (e.g., newspaper, travel section articles)	Medium	Medium	Medium
Autonomous New and popular culture: documentaries, reports, news stories, movies, television programmes	High	Medium/High	Indirect
Unsolicited Organic Unsolicited information received from friends and relatives	Medium	Low	Indirect
Solicited Organic Solicited information received from friends and relatives	High	Low	Indirect
Organic Actual visitation	High	—	Indirect

Source: Gartner 1993

recover more quickly from the image formation effects of negative news coverage (for more information about this, see Milo and Yoder 1991).

Understanding how the image formation process affects various groups is another important consideration. If the travel decision-making body is comprised of a family unit, induced image formation agents may be a top priority. Gitelson and Crompton (1983) found family groups were more likely to use media sources for information acquisition than singles. They also found that college-educated individuals were more likely to use Destination Specific Travel Literature (e.g. guide books and brochures). Capella and Greco (1987) found that people over sixty years of age were more inclined to rely on solicited organic image formation agents as this group's destination selection was greatly influenced by families and friends. Some print media (i.e. magazines and newspapers) were also important information sources for this group. Age was also found to be a factor in determining credibility ratings of various companies (Weaver and McCleary 1984) and further investigation may show a relationship between age and different types of image formation agents.

Timing also has to be considered. Van Raaij and Francken (1984) found Overt Induced I sources of image formation to be important information sources early in the decision process. Overt Induced II sources enter into the information search process at a later stage. Throughout the information acquisition period, solicited organic sources were used at the same rate.

Finally, the type of image(s) to be projected must be addressed. If a strong brand image already exists, less money and effort will be required to develop a local area image which is consistent with the dominant brand image. On the other hand, if a strong unique image which is independent or counter to the prevailing brand image is to be projected, all image formation agents should be considered as important contributors. The amount of money required for a small entity to establish an image different from the prevailing brand image requires large numbers of tourists be hosted to justify the expense and time involved in forming a unique image. Consequently, mass tourism markets will have to be developed and the issue of sustainable tourism systems becomes even more important.

Small-scale tourism developments, using alternative forms of tourism, would be better off to avoid widespread use of induced formation agents and to rely primarily on the organic types to develop their touristic image. In this way the type of images formed will be consistent with the type of experiences offered.

CONCLUSIONS

Initially, it would appear that there is a direct relationship between how images are formed and sustainable tourism systems. Hopefully, after reviewing the material in this chapter the reader will conclude that this is not the case. If it is accepted that images are formed in many different ways, and that these differences can be categorized and differences between categories identified, then it is possible to select the right image formation mix to achieve a certain size of tourism development. Obviously there are many factors that must be taken into consideration for the above statement to be true, including such diverse factors as who controls the development process – a much more complex question and one that space does not permit addressing adequately here. The point is that the image formation process is simply a method that when used properly can help achieve sustainable tourism systems, and when used improperly can hasten the onset of serious problems. The key is to define sustainable. Once defined, all the planning, development and marketing tools and methods are required to achieve it. The image formation process is one of the tools that can, and should, be used.

REFERENCES

Atkin, C. and Block, M. (1983) 'Effectiveness of celebrity endorsers', *Journal of Advertising Research*, 23, Feb/Mar, 57–61.

Bitner, J. and Booms, H. (1982) 'Trends in travel and tourism marketing: the changing structure of distribution channels', *Journal of Travel Research*, 20 (4): 39–44.

Boulding, K. (1956) *The Image-Knowledge in Life and Society*, Ann Arbor: The University of Michigan Press.

Campbell, B. (1969) 'The existence of evoked set and determinants of its magnitude in brand choice behavior', unpublished doctoral dissertation, Columbia University.

Capella, L. and Greco, G. (1987) 'Information sources of elderly for vacation decisions', *Annals of Tourism Research*, 6 (4): 148–51.

Cohen, E. (1978) 'The impact of tourism on the physical environment', *Annals of Tourism Research*, 5 (2): 215–37.

Crompton, J. (1979) 'An assessment of the image of Mexico as a vacation destination and the influence of geographical location upon that image', *Journal of Travel Research*, 17 (4): 18–23.

Crompton, J. and Lamb, C. (1986) *Marketing Government and Social Services*, New York: Wiley.

Cumings, B. (1983) 'Korean–American relations: a century of contact and thirty-five years of intimacy', in W. J. Cohen (ed.) *New Frontiers in American–East Asian Relations,*, New York: Columbia University Press, 237.

Dragicevic, M. (1991) 'Towards sustainable development', *Proceedings of the 1991 AIEST Conference*, St Gall, Switzerland: AIEST, 29–62.

Friedman, H. and Friedman, L. (1979) 'Endorser effectiveness by product type', *Journal of Advertising Research*, 19, Oct/Nov, 63–71.

Gartner, W. (1989) 'Tourism image: attribute measurement of state tourism products using multidimensional scaling techniques', *Journal of Travel Research*, 28 (2): 16–20.

——(1993) 'The image formation process', *Journal of Travel and Tourism Marketing*, 2 (2/3): 191–216.

——(1996) *Tourism Development: Principles, Processes, and Policies*, New York: Van Nostrand Reinhold.

Gartner, W. and Hunt, D. (1987) 'An analysis of state image change over a twelve-year period (1971–1983)', *Journal of Travel Research*, 26 (2): 15–19.

Gartner, W., and Shen, J. (1992) 'The impact of Tiananmen Square on China's tourism image', *Journal of Travel Research*, 30 (4): 47–52.

Gitelson, R. and Crompton, J. (1983) 'The planning horizon and sources of information used by pleasure vacationers', *Journal of Travel Research*, 21 (3): 2–7.

Goodrich, J. (1978) 'A new approach to image analysis through multi-dimensional scaling', *Journal of Travel Research*, 16: 10–13.

Grass, R. and Wallace, W. (1969) 'Satiation effects of television commercials', *Journal of Advertising Research*, 9, September, 3–8.

Gunn, C. (1972) *Vacationscape: Designing Tourist Regions*, Austin TX: Bureau of Business Research, University of Texas.

Gyte, D. (1988) *Tourist Cognition of Destination: An Exploration of Techniques of Measurement and Representation of Images of Tunisia*, Nottingham: Department of Geography, Trent Polytechnic.

Howard, J. (1963) *Marketing Management*, Homewood IL: Irwin Publishing Company.

Hunt, J. (1971) 'Image: a factor in tourism', unpublished doctoral dissertation, Fort Collins CO: Colorado State University.

Inskeep, E. (1991) *Tourism Planning: An Integrated and Sustainable Development Approach*, New York: Van Nostrand Reinhold.

International Union for Conservation of Nature and Natural Resources (1980) *World Conservation Strategy: Living Resource Conservation for Sustainable Development*, Gland, Switzerland: IUCN.

Khan, S. (1991) 'Nonresidents perceptions of Wisconsin's tourism regions', unpublished MS thesis, Menomonie WI: University of Wisconsin-Stout.

Kotler, P. (1982) *Marketing for Non Profit Organizations,*, Englewood Cliffs NJ: Prentice Hall.

McLellan, R. and Noe, F. (1983) 'Source of information and types of messages useful to international tour operators', *Journal of Travel Research*, 8 (3): 27–30.

Milo, K. and Yoder, S. (1991) 'Recovery from natural disaster: travel writers and tourist destinations', *Journal of Travel Research*, 30(1): 36–39.

Murphy, P. (1983) 'Perception of attitudes of decision making groups in tourist centers', in M. Stabler (ed.) *The Image of Destination Regions*, New York: Croom Helm.

Narayana, C., and Markin, R. (1975) 'Consumer behavior and product performance: an alternative conceptualization', *Journal of Marketing*, 39: 1–6.

Okoroafo, S. (1989) 'Branding in tourism', in S. F. Witt (ed.) *Tourism Marketing and Management Handbook*, New York: Prentice Hall, 23–6.

Pacific Travel News (1984) 'Evaluating Korea', February, 38–40.

Pearce, D. and Turner, R. (1990) *Economics of Natural Resources and the Environment*, New York: Harvester Wheatsheaf (cited in Dragicevic 1991).

Phelps, A. (1986) 'Holiday destination image: the problem of assessment; an example developed in Menorca', *Tourism Management*, 7 (3): 168–180.

Roehl, W. (1990) 'Travel agents' attitudes toward China after Tiananmen Square', *Journal of Travel Research*, 29 (2): 16–22.

Schumann, D., Petty, R. and Clemons, S. (1990) 'Predicting the effectiveness of different strategies of advertising variation: a test of the repetition-variation hypotheses', *Journal of Consumer Research*, 17, September, 192–202.

Shen, J. (1989) 'Tourism image of China as perceived by the American mature traveler', unpublished MS thesis, Menomonie WI: University of Wisconsin–Stout.

Stutman, R. and Newell, S. (1984) 'Beliefs versus values: salient beliefs in designing a persuasive message', *The Western Journal of Speech Communication*, 48, Fall, 362–72.

Thompson, J. and Cooper, P. (1979) 'Attitudinal evidence on the limited size of evoked set of travel destinations', *Journal of Travel Research*, 17 (3): 23–5.

Thurstone, L. (1967) 'The measurement of social attitudes', in M. Fishbein (ed.) *Readings in Attitude Theory and Measurement*, New York: Wiley.

United Nations (1987) *Our Common Future: World Commission on Environment and Development*, Oxford: Oxford University Press.

United States Travel Service (1977) *International Market Reviews of Selected Major Tourism Generating Countries*, Washington DC: US Department of Commerce.

van Raaij, W. and Francken, D. (1984) 'Vacation decision, activities and satisfactions', *Annals of Tourism Research*, 11 (1): 101–12.

Weaver, P., and McCleary, K., 1984. 'A market segmentation study to determine the appropriate ad/model format for travel advertising', *Journal of Travel Research*, 23 (1): 12–16.

Woodside, A. and Sherrell, D. (1977) 'Traveler evoked set, inept set, and inert sets of vacation destinations', *Journal of Travel Research*, 16(1): 14–18.

Woodside, A., and Ronkainen, I. (1980) 'Vacation travel planning segments: self planning vs users of motor clubs and travel agents', *Annals of Tourism Research*, 7 (3): 385–94.

Part IV

OPPORTUNITIES AND CHALLENGES IN SUSTAINABLE TOURISM

12

NATURE TOURISM DEVELOPMENT

Private property and public use

Turgut Var

After undergoing a very serious economic downturn due to oil and banking crises in the eighties, the state of Texas started the last decade of the century with an emphasis on economic diversification and growth. Two major external factors have contributed to the economic revival of the state. Signing of the North America Free Trade Agreement (NAFTA) led to a rapid growth of exports from Texas to Mexico. Since 1987, Texas exports to Mexico have increased near or above double-digit rates, rising from about $1 billion in 1987 to about $26 billion in 1993. Major state exports to Mexico include computers and electronic equipment, industrial machinery, transportation equipment, primary and fabricated metals, and chemicals. In the future, the implementation of the North American Free Trade Agreement is expected to strengthen further the economic ties between Texas and Mexico.

The second external factor was the lower interest rate policy of the Federal Government, in large part designed to stimulate the US economy out of a recession. In 1991 the Federal Reserve Board began dropping interest rates, and continued to do so until these reached levels never before seen by most baby boomers, and this group of housing consumers acted with a vengeance by refinancing existing mortgages and purchasing new homes (Sharp 1995).

The surge in exports to Mexico preceding the fall of the peso, the interest rate fuelled construction boom and strong US industrial growth led to expansion of production in the manufacturing sector. Texas is expected to perform better than the US average between 1995 and 1997. However, the state still has the distinction of having some of the poorest regions in the country, especially along the Mexican border. Texas ranks 47th among the states in poverty rate, 40th in income distribution, 30th in rural/urban disparity and 49th in hazardous waste generation. In addition to the chronically depressed border areas, several counties have reacted to the declining importance of oil and gas resources by seeking new alternatives like tourism in order to achieve higher employment and income (Corporation for Enterprise Development 1996).

In search of economic diversification Texas turned to tourism as a tool of economic development both at regional and state levels. Being in predominantly private sector land ownership (only 3 per cent of land in Texas is public land), Texas had to initiate planning activities that required coordination of various state involvements related to tourism. Under the leadership of the Department of Commerce annual plans were developed. These plans stressed the importance of preservation of historical, cultural and natural resources of the state in developing tourism to raise the welfare of Texans. Since the role of the state has been catalytic rather than direct in the private sector dominated economy most of the actions have required persuasion and a grassroots approach. Among the alternatives available, nature-based tourism seemed to be one of the attractive avenues of tourism development. The objective of this paper is to give a short history of nature-based tourism development in Texas and discuss its future as an economic development tool.

According to most recent statistics tourism represents more than $25 billion in business for Texas. In 1994, a record 166 million domestic visitors travelled to Texas, solidifying the state's ranking as the second most-visited state after California and ahead of Florida (State Task Force on Texas Nature Tourism 1994). Tourism in Texas supports close to 435,000 jobs and an $8 billion payroll (Texas Parks and Wildlife Department 1996). An important component of this sector is nature tourism, itself one of the fastest-growing segments of global travel. Eight of the top twenty reasons non-Texans vacation in Texas are related to nature. These include, beautiful scenery, attractive beaches, state parks, lakes and boating, fresh and saltwater fishing, good campgrounds, good hiking trails and dude ranches. According to a Texas Parks and Wildlife Department study, visitors to state parks spent $179 million in 1993, with an estimated total economic impact of $477 million. An estimated $3.6 billion was spent on recreation associated with fishing, hunting and wildlife during the same year. A report prepared by the US Forest Service suggests that backpacking will grow by 34 per cent through the year 2000. Similar growth rates from 31 per cent to 11 per cent are forecast for day hiking, bicycling, outdoor photography, wildlife watching, camping, canoeing/kayaking, and rafting/tubing (STFTNT 1994).

Recognizing the opportunity, a special State Task Force appointed by the Governor convened in late 1993 to develop a report on nature tourism. The Task Force was given the following mission (STFTNT 1994: 2):

1 Examine the potential of nature tourism in Texas.
2 Recommend opportunities for developing and promoting it.
3 Build upon local efforts already under way.
4 Preserve local, social and cultural values.
5 Promote sustainable economic growth, restorative economic development and environmental conservation through nature tourism.

One of the first duties of this Task Force was to define nature tourism. According to the Task Force nature tourism is discretionary travel to natural areas that conserves the environmental, social and cultural values while generating an economic benefit to the local community. Similarly, nature tourists are travellers who spend their time and money enjoying and appreciating a broad range of outdoor activities that have minimal impact on the environment. The Task Force completed its mission and submitted its findings in a report named *Nature Tourism in the Lone Star State: Economic Opportunities in Nature*. The report contains several recommendations grouped into four categories: conservation, education, legislation and promotion. These are discussed below.

CONSERVATION

Conservation is a very touchy issue in Texas, a state in which 97 per cent of the land is owned by the private sector. Known for its ecological diversity, Texas is the home of 5,500 plant species, 425 of which only occur in Texas. Of 1,100 vertebrates in Texas, 60 are found nowhere else in the world. With 540 bird species, Texas has more than any other state in the US (Texas Center of Policy Studies 1995).

Judging by the attention that US environmental protection laws are being given today, it would seem that the country's focus on wildlife protection has only just begun. Nothing could be further from the truth. Faced with diminishing wildlife species due to unregulated hunting and to the 'civilizing' of the frontier, the United States began legislatively protecting wildlife and habitats in the early nineteenth century. Texas also was active as far back as 1830 when a Fish Commissioner was appointed. For the most part these legislative actions and regulations governing wildlife, treat animals, birds and fish as harvestable resources, much like trees: only so much can be taken at one time, leaving just enough to continue the reproduction process. Some of the significant legislation at federal level included the National Park Service Organic Act of 1916, the Migratory Bird Act of 1918, the Fish and Wildlife Coordination Act of 1958, and the Marine Mammal Protection Act of 1972 which banned the killing and importing of whales and nearly all other marine mammals.

However, many authorities believe that the most far-reaching legislation has been the Federal Endangered Species Act of 1973. Though more than twenty-three years old, this legislation remains controversial in many states, especially Texas.

In enacting the Endangered Species Act, Congress addressed the question of why money should be spent to save non-human species. In the preamble to the Endangered Species Act, the US Congress states that 'species of fish, wildlife and plants are of aesthetic, ecological, educational, historical, recreational and scientific value to the Nation and its people'. According to the Texas Center for Policy Studies (1995) the key elements of the Endangered Species Act are:

1 Congress requires the US Fish and Wildlife Service (USFWS) to devise a process of listing species that are endangered or threatened.
2 USFWS must identify critical habitat of the species.
3 The Act also sets out requirements that prohibit the taking, possession, transportation or sale of any species designated as threatened or endangered without a permit.
4 Any federal agency whose actions might affect an endangered species or its habitat must consult with USFWS.
5 Non-federal actions require approval of USFWS where there is a 'take' of endangered species, provided such 'taking' is incidental to and not the purpose of activity. Take under the Endangered Species Act includes killing, capturing, harming and harassing. Harm includes modifying habitat to the point that the species' breeding, feeding or sheltering is impaired.

Following the federal legislation in 1973 the Texas Legislature enacted a state Endangered Species Act, which was amended several times (Texas Center for Policy Studies 1995). The Act gave the Texas Parks and Wildlife Department (TPWD) authority to establish a list of fish, wildlife and plants endangered or threatened with statewide extinction. Once listed, these species are given protection and are eligible for restoration efforts.

In the Endangered Species Act (ESA), there is no state criterion for protecting wildlife species from indirect take (e.g. destruction of habitat or unfavourable management practices). The TPWD does have a Memorandum of Understanding with every state agency to conduct a thorough environmental review of state-initiated and state-funded projects such as highways, reservoirs, land acquisition and building construction, to determine the potential impact on the state's endangered or threatened species.

It is interesting to note that in 1991 the Texas Legislature adopted a bill adding wildlife management to the list of *agricultural uses* for property tax purposes. According to this bill the land retains its agricultural-use property tax status (Texas Center For Policy Studies 1995). This issue will be dealt with in the legislation section of this paper.

The Endangered Species Act has received a number of criticisms from both landowners and environmentalists. Environmentalists claim that the Act has treated plant species differently from wildlife. Plants are considered to be part of the real estate on which they grow, and thus they are legally treated as part of private property. Animals, however, are part of the public trust and not owned by the property owner. They also argue that high-profile species receive more funding for recovery while invertebrates and plants disappear. Some of them even suggest that the Act is a crisis-based approach to species protection because it was designed to save species already on the verge of extinction, rather than foster a strategy that protects habitat (Texas Center for Policy Studies 1995).

The private landowners, on the other hand, argue that the Endangered Species Act has burdened them with the major responsibility of protecting endangered species without any compensation. They claim that the Act lowers property values, and because of a cumbersome permit process, their ability to sell their property is hindered. Furthermore, some landowners assert that the Act is an assault on Fifth Amendment private property rights.

According to the Center for Policy Studies, the Act has become the target for special interest debate when in fact it should be a common interest debate. Both conservationists and landowners do seem to agree that species – not just endangered species – and habitat protection is important and cannot be achieved solely through government regulations.

Considering the above background, the Task Force for Nature Tourism (1994) recommended a four-pronged conservation strategy:

1 Provide incentives to private landowners to preserve natural habitats.
2 Manage public lands, such as state parks and wildlife management areas, for the enrichment and continuance of wildlife diversity.
3 Lease or acquire additional lands from willing sellers where TPWD is able to manage, enhance and conserve habitats for all wildlife diversity and to provide for a wide range of recreational opportunities.
4 Better utilize mitigation funds for acquiring and enhancing lands that may additionally function as nature tourism destinations.

The Task Force also recommended that private and public sector efforts in nature tourism and resource conservation activities be coordinated. Local, state, federal and private sector organizations and companies should be identified according to current areas of responsibility and potential needs. In order to achieve this objective, a memorandum of understanding or inter-agency agreement should be developed between related agencies, and an inter-agency/private task force should be established to facilitate nature tourism conservation activities (STFTNT 1994: 17).

EDUCATION

As mentioned above, Texas is experiencing a fundamental economic restructuring after the oil and real estate crises of the 1980s and early 1990s. Most of the state's population lives in five large cities, namely Houston, Dallas, Fort Worth, Austin and San Antonio. According to a report from Texas Rural Communities Incorporated, only 18 per cent of the state's population resides in 204 rural counties. Most of these counties, especially those in western Texas, have experienced important population losses, with many communities losing up to one-third of their population in the last ten years. All these communities are trying to increase their share in state tourism in order to diversify their economies (STFTNT 1994).

Recognizing the need for training rural community leaders in tourism development, the Task Force has made several suggestions related to education. These are as follows:

1 Develop a step-by-step nature tourism handbook targeted to communities and private landowners.
2 Provide training and outreach for local communities, individuals and companies to nurture and enhance nature tourism in their areas.
3 Enable the development of local tourism infrastructure to support the nature consumer's needs.
4 Provide training for public and private sector employees who interact with the public concerning basic hospitality skills and nature tourism opportunities in their areas.
5 Identify and coordinate public and private organizations with the financial resources and expertise to help communities and individuals in their nature tourism efforts.
6 Identify nature tourism products and infrastructure which are both available and needed to promote sustainable growth and environmental conservation.
7 Develop programmes to communicate the importance of protecting and managing the state's nature resources.

The above recommendations do not give the responsibility for education to one agency. They emphasize the importance of cooperation and coordination of various public and private activities. Though nature tourism holds the promise of economic development for local communities, it has also the potential to damage the very resource that these communities mean to benefit from.

LEGISLATION

The recommendations of the Task Force on legislation can be summed up in four main areas:

1 Property tax relief for lands devoted to wildlife management and nature tourism.
2 Incentives for transportation companies serving rural areas.
3 Limitation of liability and removal of the cap on entrance fee revenues.
4 Federal probate relief for landowners who manage their land for wildlife habitat and outdoor recreation.

Currently, in order to receive relief from *ad valorem* property taxes, landowners must operate their farms and ranches for agricultural purposes. This requirement discourages landowners who desire to establish a nature tourism business but cannot afford to take the risk of losing the relief from *ad valorem* agricultural taxes.

Nature tourism requires a good transportation network that would make the experience unique. In most cases the resource is located in those areas without proper transportation service due to low profitability. The Task Force felt that in order to develop nature tourism a serious overhaul of the existing transportation laws is necessary. The new legislation can create an incentive for transportation companies to operate in rural regions of the state (STFTNT 1994).

Many Texas landowners are hesitant to allow public access because of liability exposure. It is necessary to develop an insurance programme for landowners who are interested in providing for nature tourism on their lands. Finally, landowners presently are granted a limitation regarding recreational activities as long as revenues received do not exceed twice the previous calendar year's *ad valorem* tax. The Task Force strongly argued that rural landowners engaged in nature tourism should have the same tax advantages allowed for farming and ranching, and recommended the removal of this cap.

Another point is related to federal probate taxes. The average age of rural Texas residents is ten years older than the median age for the state, and the average age of farmers and ranchers, who manage the largest land-asset base, is nearly sixty. As the descendants of the farmers and ranchers inherit the land, many are forced to sell or subdivide the property. This fragmentation has diminished the contiguous wildlife acreage in the state. The State Task Force on Nature Tourism recommended that one form of relief from the federal probate tax, through tax credits in exchange for conservation easements, would lessen the fragmentation and increase the area for wildlife habitat.

PROMOTION

In order to achieve the full benefit of developing nature tourism destinations, the Task Force recommended formation of the Texas Nature Tourism Association, which would be responsible for issuing voluntary guidelines for operators, providers and sites that wish to be certified for quality nature tourism. It further suggested that the Texas Nature Tourism Association should be a not-for-profit industry organization composed of companies and individuals with an interest in nature tourism. Several responsibilities, including development of a handbook, promotion and competition in the market place, were recommended for this prospective association (STFTNT 1994).

In spite of a change in government most of the recommendations of the Task Force were implemented. The last portion of this paper deals with recent developments in nature tourism in Texas.

Texas Nature Tourism Association

In early 1996 the Texas Nature Tourism Association was formed as a not-for-profit organization. The main support for this organization came from the

Texas Travel Industry Association. According to the bylaws of the Association the President shall be nominated and hired by the President of the Texas Travel Industry Association. In addition to the appointment of the President, one member of the Board of Directors is appointed by the Texas Travel Industry Association. The Texas Nature Tourism Association pays a management fee to the Texas Travel Industry Association, the stated purpose of which is:

> education, research and information exchange to encourage the environmentally sound use of the natural landscape of Texas including the land, water, plant and animal resources. Through research and demonstration the corporation will develop examples and methods to perpetually care for the natural environs of Texas and will provide to the overseers of public lands and the owners of private lands information regarding methods to preserve, restore, maintain and make accessible those areas for the public's use and enjoyment.
>
> (Texas Nature Tourism Association 1996: 1)

It is also interesting to note that as a not-for-profit organization, the Association is restricted by its bylaws not to attempt to influence legislation nor engage in any political campaign on behalf of any candidate for public office.

Making nature your business – a guide for starting a nature tourism business in the Lone Star State

In line with the recommendations of the Task Force, the Texas Parks and Wildlife Department, Texas A&M University and the Lower Colorado River Authority came up with a working document called *Making Nature Your Business*. This thirty-six page booklet contains interesting sections for those that would like to engage in nature tourism. It clearly describes nature tourism and tourists and supplies several success stories, including the Great Texas Coastal Birding Trail soon to become a reality. It contains three sections that are very important for nature tourism. These are:

1 Guidance for natural and cultural resource managers and community leaders.
2 Guidance for a quality nature tourism industry.
3 Guidance for tour guides and nature travellers.

The first set of planning guidelines stresses carrying capacity and the impacts of nature-based tourism on the resource, the community and the tourism industry. The second set of guidelines emphasizes that nature-based tourism businesses should try to work closely with local community leaders and resource managers to plan for nature travel experiences compatible with community and resource management goals. Tourism should contribute positively to local communities and resources. Businesses are expected to adopt or reaffirm an environmental stewardship ethic for their activities. Finally, guidelines for tour guides and nature travellers are given in point form and include planning, camping ethics, safety in the outdoors and behaviour in the wilderness.

The rest of the book contains basic principles of planning, development, marketing, operations and management. The sources of information are also presented at the end of book. As a working document, *Making Nature Your Business* is intended to provide a user-friendly and readable resource for persons interested in starting successful nature tourism projects that are sensitive to the natural resources on which a nature business depends (Texas Parks and Wildlife Department 1996).

Nature tourism in practice

Interest in nature tourism has been reflected in several festivals related to wildlife. In addition to traditional hummingbird festivals and bird watching on coastal areas of Texas, other significant nature-based birding festivals that have developed in recent times include: Crane Fest, Prairie Chicken Festival, Spring Migration Celebration, Bluebird Festival and Eagle Fest.

The Greater Texas Coastal Birding Trail, when completed, will make it easy for bird watchers to drive from one end of the Texas coastline to the Mexican border in the south. Highlights of the trail include nine national wildlife refuges managed by the US Fish and Wildlife Service and many Texas Parks and Wildlife management areas and sites that offer unique birding experiences. The project is the result of the cooperation of public and local community organizations. Each site along the 500-mile coastal highway trail will be marked with a special sign and shown on a map in an interpretive guide developed for the trail. Many sites will be enhanced and expanded to increase attractiveness of wildlife habitat to both birds and bird watchers (Texas Parks and Wildlife Department 1996).

CONCLUSIONS

The literature on conservation and development disputes is extremely rich. Recently the US Congress has engaged a number of hearings on the question of takings and the implementation of the Wild Bird Conservation Act of 1992. In all these discussions private property rights and infringement of these rights in creating or expanding natural habitat for wildlife have occupied centre stage. In a state where land belongs predominantly to the private sector, the solution is very difficult because both parties have valid arguments. It is obvious that legislation alone cannot provide an answer, and can cause more divisions among the interested parties. Voluntary cooperation seems the most logical approach in many development problems. The case of nature tourism for Texas is a good example of private and public cooperation in achieving a solution both for conservation and public recreational use of private property.

REFERENCES

Corporation for Enterprise Development (1996) *News Release*, Washington DC, 11 July.

Sharp, J. (1995) *Gaining Ground: A Regional Outlook*, Austin TX: Texas Comptroller of Public Accounts.

State Task Force on Texas Nature Tourism (1994) *Nature Tourism in the Lone Star State*, Austin TX: Texas Parks and Wildlife Department.

Texas Center for Policy Studies (1995) *Texas Environmental Almanac*, Austin, TX: Texas Center for Policy Studies.

Texas Nature Tourism Association (1996) *By-laws*, Austin TX: Texas Nature Tourism Association.

Texas Parks and Wildlife Department (1996) *Making Nature Your Business*, Austin TX: Texas Parks and Wildlife Department.

13

CULTURAL AND LANDSCAPE TOURISM

Facilitating meaning

Richard Prentice

INTRODUCTION

This chapter reflects the changed role which landscape and culture represent to 'western' societies as the new millennium is approached. No longer are such resources largely to sustain local inhabitants, they are now the focus of consumption by consumers resident elsewhere. In response, tourism policy now pervades the more traditional areas of social, economic, environmental, rural and urban policies (e.g. Prentice 1993a). As such, cultural and landscape tourism implies an ultimacy of consumerism: the legitimacy and ability, although temporarily and often imperfectly, to choose as easily as selecting a new sound system or wall hanging, another culture or landscape as a setting for experience. As illusion, outwith the tourism ghettos of non-place differentiated tourism, the tourism settings provided by culture and landscape are a hallmark of postmodernity, since for many such transposition is no longer an extravagance but is perceived instead as a 'right' of modern affluence. Residential relocation to areas selected for their cultural or landscape value is the final manifestation of this 'right' and achievement of illusion; relocations are frequently the consequence of earlier tourist experiences.

In view of this hallmark of contemporary western living, the present focus is on the *consumers* of cultural and landscape 'products' as tourists. The discussion originates from the belief that a sustainable product needs to be sustainable simultaneously in two ways and to two groups; namely in production and in consumption, and to destination communities and to tourists. This is a broader definition of sustainability than often used, definitions more usually focusing at least implicitly on host communities, that of tourism in harmony with its physical, social and cultural environment (*cf.* Medlik 1993). In contrast, the view underlying this chapter is that only part of sustainable production and consumption is borne by those communities receiving tourists. This is because tourists themselves contribute through their imaginings not just to consumption, but to the production of their own cultural (and often to their own landscape) tourism 'product'.

Because of this, in cultural tourism, differentiation between production and consumption is far from distinct.

The effects of tourism on the communities and physical environments receiving tourists are far more widely known than the effects of tourism on tourists, and how beneficial effects may be facilitated through cultural and landscape imagining. In consequence this chapter seeks explicitly to redress this balance, and offers a largely demand-side analysis. Demands for cultural and landscape tourism are reviewed, with particular attention paid to environmental preference and demands for insight into what is being viewed or otherwise experienced. A generic framework for understanding cultural and landscape tourism consumption is offered in the form of a *multiple consumption matrix*.

CULTURAL AND LANDSCAPE TOURISM IN THE CONTEXT OF SUSTAINABILITY

Cultural and landscape tourism is about the exploitation and enjoyment of cultures and landscapes as forms of tourism. As such, cultural and landscape tourism should not be equated with the exploitation of fine art or 'high' culture, as this is to ignore the popular aspects of such exploitation and enjoyment. Cultural and landscape tourism is a much broader concept than a focus on palaces, cathedrals, temples and national galleries might imply. Its resources include those of historical geography, archaeology, literature and environmental management, to name but a few (*cf.* Prentice 1993a, 1993b). In essence, cultural and landscape tourism is about what a geographer would term *place*, the understanding of 'places as they really are' (Robinson and McCarroll 1990: 3), and about *heritage*, things used as tourism *place-products*, 'which are literally or metaphorically passed on from one generation to the other' (Prentice 1993b: 5).

At the same time cultural and landscape tourism is a form of self-initiated expressive behaviour, but one which has for long been seen as potentially disruptive, if not destructive, of the places visited (Mathieson and Wall 1982) and which is increasingly being seen as sub-optimal through tourist volumes and organization (Glasson *et al.* 1995). Cultural and landscape tourism is also about goods and services which are one and the same time both unnecessary and necessary to life in contemporary western society. It is one aspect of what the Henley Centre (1996) has termed 'mobility hunger' among contemporary western populations. The justification to its consumers of cultural and landscape tourism may be found in Graburn's description of tourism as 'functionally and symbolically equivalent to other institutions that humans use to embellish and add meaning to their lives' (1978: 17). This is a conceptualization of consumption as a mix of internalized meanings and external symbolism (*cf.* McCrone *et al.* 1995) and is important in that demand can only be understood fully in terms of what the consumer is seeking to get out of their visit directly, and indirectly through the reactions of others in their home community as symbolic additional value. The present chapter concentrates on the understanding of the former, namely what tourists are directly seeking. This is simply because the extent of indirect additional value added through cultural and landscape tourism still awaits extensive research attention, whereas the meaning of tourism has attracted at least intermittent interest in research.

The conceptualization of cultural and landscape tourism as a form of self-initiated expressive behaviour implies important parallels with other forms of this behaviour, 'outwith tourism'. As a popular activity it is akin, for example, to embroidery, lacemaking, modelling or flower gardening as a popular cultural expression (cf. Ó Cléirigh 1985; Garrad and Hayhurst 1988; Hendry 1994; Bussi 1995; Scarman 1996). Like these other activities, cultural and landscape tourism allows the demonstration of creativity through imagining, through the 'production' of artefacts to reproduce experience in tangible form (in the case of tourism, frequently through photographs and videos) and in the collation of artefacts associated with the activity (in this case, souvenirs). As for these other activities, a motivation of cultural and landscape tourism is the inherent pleasure of the process; a motivation which may be achieved to different skill levels and with different investments of time and money. It is also an activity which may be perceived by consumers as personal to the individual or family, and therefore implicitly unharmful to others.

Tourism is rapidly generating a plethora of definitions. Those pertaining to cultural and landscape tourism as a sustainable form of production and consumption are no exception. In consequence, some attention is needed to terminology, and an awareness that terms such as *alternative* tourism, *ecotourism* or *ethnic* tourism are inappropriate generic descriptions for tourism involving cultural or landscape appreciation. Each of these definitions essentially focuses on specialist markets. As the resource of place is central to cultural and landscape tourism and is widely demanded as at least a setting for experiences, *alternative* is an inappropriate generic description for cultural and landscape tourism. Such appreciation is mainstream; likewise the sustaining of this resource needs to be mainstream if generic demands are to be met. Landscape tourism is also broader than the specialist product term *ecotourism* (nature-based tourism) implies; fauna and flora can be settings for other experiences as well as the foci of experiences. Indeed, some recent definitions of ecotourism have effectively broadened the definition of this aspect of landscape to include 'cultural components' within the term 'natural environment' (Commonwealth Department of Tourism 1994; Cater and Lowman 1994) thereby embracing cultural tourism within the focus of a wider ecological tourism to 'primitive' places. Indeed, the implicit equation of ecotourism with remote or 'primitive' places renders this term deficient as a generic conceptualization of landscape tourism, as European settled landscapes are implicitly excluded, but are at the same time the destination of many tourists seeking landscape. Not only are many European settled landscapes resultant of human interaction with the wider environment, they are now actively managed as much for their resultant appearance as for other benefits (see for example Countryside Council for Wales 1995a, 1995b, 1995c, 1996). An equation of landscape tourism with ecotourism is also deficient as first-hand experience of the natural environment pervades discussions of this form of nature consumption, with consumers seen essentially as outdoor enthusiasts (Ballantine and Eagles 1994; Johnston and Edwards 1994). Such enthusiasts are part only of demand for landscape appreciation and the recognition of differing ecotourism demands is essential (e.g. Orams 1995). Finally, *ethnic* tourism in either of its forms as visiting quaint places or travel for ethnic reunion (King 1994) is equally too restrictive a conceptualization of cultural tourism, for it excludes city travel other than for purposes of ethnic reunion, for example.

In contrast to how cultural and landscape tourism may be perceived by its participants, that cultural tourism can effect cultural change in destinations, and likewise landscape tourism can effect environmental change, has been a common focus in calls for sustainable development. This is because, unlike many other industries, consumption as well as production externalities need to concern policy makers in countries supplying tourism products (Forsyth *et al.* 1995). Examples of management response abound. Initiatives have ranged from international and national strategies (e.g. Commonwealth Department of Tourism 1994; Mason 1994; Meldon and Walsh 1995) to local management plans, including those for cities and national parks. Scotland, for example, has a multi-agency *Tourism and Environment Initiative* (Tourism and Environment Task Force 1996) as well as integrated environmental tourism, retailing arts and crafts, and housing initiatives in Edinburgh (Parlett *et al.* 1995). City initiatives are now common across Europe as tourist numbers increase (e.g. Glasson *et al.* 1995). Such initiatives have been variously characterized as technical and imprecise, valuable if they are better than counter-productive rather than optimal (Forsyth *et al.* 1995; Hughes 1995). Absent from many city strategies has been explicit *experience or benefits-based management*, whereby the experiences or benefits gained by consumers or thought appropriate by policy makers to induce become a focus of facilitation through management.

Comparable attention to that given to impacts has not been given to the extent to which cultural and landscape 'consumption' affects short-term or lasting change to tourists. This is despite the environmental message implicit in much ecotourism debate, and the assumption that tourists' behaviour can and should be modified to respect the environment. A substantial exception to this neglect is to be found in North American outdoor recreation research, which has addressed the benefits sought and gained through activity holidaying, and sought to apply these in experience and benefits-based management (Pigram 1993). Not only is it unlikely that tourist volumes will be diverted successfully or otherwise controlled if demand as motivation is not properly understood, to neglect the experience of tourists implicitly undervalues tourism. This neglect also fails to address the issue of how far tourism may be sustainable for tourists, either as repeat visitors to the same place, as recommendation to others, or as repeat tourists visiting elsewhere. The latter issue includes that of tourist 'careers' or 'bibliographies' (Cohen 1979a; Pearce 1993; Ryan 1995), a concept which warrants greater attention than so far given across different types of tourism and tourists. Knowledge of this sort is essential to informing tourism development and management. As Pigram has commented for rural tourism,

> tourist settings and experiences do not occur spontaneously. . . . Creation of an appropriate range of settings for rural tourism requires the deliberate selection and manipulation of features of the rural landscape to accommodate different types and styles of visitor use.
>
> (Pigram 1993: 163)

CHARACTERIZATION OF DEMAND

Demand for cultural and landscape heritage may be characterized as diverse, but predominantly generalist. At the level of the individual consumer, 'aesthetic cosmopolitanism' provides a potentially useful generic conceptualization of demand; at the level of international policy, drivers for cultural integration and tolerance may be found as bases for the orchestration of demands, supplementing economic drivers in the forms of economic development and restructuring.

Generalist consumer demands

Demands for cultural and landscape tourism cannot sensibly be divorced from a context of wider changes in consumer demand in western societies and elsewhere. Three such changes may be identified as highly significant: increasing consumer confidence, increased growth of the 'middle' classes and scarcity of leisure time (Henley Centre 1996). Growing consumer confidence in purchasing and the widening of choice in many markets are causing a decline in preconceptions about appropriate products and suppliers, thereby increasing consumer preparedness to try new service concepts and 'product recipes'. Consumer experience in one market, where levels of service and quality are of higher-than-average standard, are increasingly being carried forward into other markets, where delivered goods or services may no longer match these raised expectations. Global experiences will increase demands for new levels of service and novelty of concepts, as consumers increasingly draw on experiences from around the world to inform attitudes and expectations. More time and money is being spent by individuals on education, thereby removing expenditure from the leisure budget. Rising educational attainment is aiding the growth of the ABC classes, the very groups disproportionately found among heritage consumers. The increase in this social group will increase the number of confident and knowledgeable consumers. At the same time, the amount of leisure time is likely to contract for these significant groups of consumers, as a result of changing working structures, mounting work pressures, increased female participation in the labour force, commuting and the need for continuous education. This is perhaps the opposite of what might have been thought ten years ago, and implies the fallacy of defining leisure as 'not-work'. These pressures are leading to increases in the value of time allocated during free hours and increased expectations about the quality of services and products. As leisure repertoires are widening, money and time are being spread more thinly across a range of leisure choices. In consequence,

> Consumers will increasingly be demanding of value for time as well as value for money. Consumers will require immediate service and products and services will increasingly be sold on a platform that promises to save time or be 'worth time'.
>
> (Henley Centre 1996: 8)

Critically, these three factors are beginning to interact together:

> Heightened expectations for leisure experiences by people who are allocating an increasingly scarce resource of time, will become ever more difficult to match. The danger that people will be 'disappointed' with the service experience will rise. Unpreparedness to be disappointed will increase as people's repertoires of choice and consumption experience broaden.
>
> (Henley Centre 1996: 38)

Until recently, pleasure and utility were usually fused in consumer theory; a fusion Campbell (1987) described as a confusion of need satisfaction with the objective of pleasure. In societies where needs are satisfied regularly and their meeting is guaranteed, the pleasures associated with overcoming discomfort are removed and individuals seek pleasures elsewhere. These changes have led van Raaij to describe the 1990s as 'a time of incessant choosing' (van Raaij 1993: 542) by consumers; and Gabriel and Lang to talk of 'the fetishization of choice' (Gabriel and Lang 1995: 46). Campbell also identified a further change: modern hedonics (pleasurable experience-seeking) has shifted its primary concern from sensations to emotions, the latter being an immensely powerful source of pleasure:

> Modern hedonism tends to be covert and self-illusory; that is to say, individuals employ their imaginative and creative powers to construct

mental images which they consume for the intrinsic pleasure they provide, a practice best described as day-dreaming or fantasising.

(Campbell 1987: 77)

Herbert has argued much the same for literary places as tourist attractions; the

attraction of such places is that they have connections not just with the life of the writer but also with the works which they created. There is a merging of the real and the imagined which gives such places a special meaning.

(Herbert 1996: 77)

The modern tourist is a dreamer in this view; a role of the tourism industry is to facilitate this imagining.

Although much contemporary consumption is primarily hedonic it should not be thought of as exclusively so. Tourist consumption can be also *instrumental* to a higher goal through gaining experiences. Reference to segmentations of tourist demand generally shows that demand for informed experiences or insight seems to be increasing worldwide, particularly among the affluent and educated middle classes. Evidence from Europe, North America and Asia confirms the generality of this demand (Prentice 1989, 1994; Richards 1994; Helber 1995). However, few of these tourists would seem to be seeking formal learning or education (Prentice 1989; Orams 1995; Herbert 1996). As such, part only of this demand is for so-called special interest tourism; much more is for culture and landscape as a background setting into which some general insight is desired while visiting. The increasing confidence of tourists through experience and education has led to a fracturing of demand. In consequence, tourists are increasingly rejecting the 'legislator' approach to tourism provision (the single authoritative statement about what is to be understood) in favour of the 'interpreter' role (the provision of alternatives) (Urry 1995). In other words, the wider expectation of choice is being applied in the tourism context.

As such, in terms both of hedonics and instrumental objectives, the facilitation of imagining may be postulated as a demand in cultural and landscape tourism. This imagining neither begins nor ends with the tourism trip. It occurs in anticipation, when present at a destination, and in memory, and potentially in virtual settings. Imagining may be seen as part of Kelly's full cycles of experience, from anticipation, through investment as immersion, through encounters as interaction of tourist and events, through confirmation and disconfirmation in assessing what has happened, to constructive revision as invigoration and enlivenment (Kelly 1955; Botterill and Crompton 1996). This invigoration and enlivenment are the ultimate generic benefits to consumers of cultural and landscape tourism.

Consumer segmentation

Contemporary marketing emphasizes segmentation, but care needs to be taken not to over-compartmentalize demand. 'Heritage' is an ambiguous word (Prentice 1993b) and reflects a blurred view of exactly what is important both to conserve and to appropriate, literally, metaphorically or by association. Culture and landscape, however, recur among explicit and implicit definitions of heritage (e.g. An Foras Forbartha 1969; Convery *et al.* 1994; Feehan and O'Donovan 1996), the latter sometimes as an aspect of 'natural' heritage (An Foras Forbartha 1985; Milton 1990). Recently, Masberg and Silverman (1996) have argued that in reality consumers are not discrete in their conceptualizations, but that history and culture recur as definitions. Salient aspects mentioned by their convenience sample of Indiana University recreational administration students included the activities undertaken, companions on the visit, site personnel, information learned, function or condition of buildings, aspects of the natural environment, and the ways people were depicted or observed. Culture and landscape (the latter as natural environment) may be identified among these elements.

Past phenomenologies of tourism have segmented demand both in terms of primary motivation and in terms of primary objective. Cohen (1979a) produced a typology of five modes of tourist experience based on motivations pertinent to cultural and landscape tourism:

- the *recreational* mode: the tourist seeking relaxation and recreation to restore their general sense of wellbeing;
- the *diversionary* mode: the tourist seeking escape from boredom, seeking to make alienation bearable;
- the *experiential* mode: the tourist looking for meaning aesthetically in the lives of others;
- the *experimental* mode: the tourist sampling alternative lifestyles in places distant from their home; and
- the *existential* mode; the tourist who has achieved enlightenment by embracing the culture of a place distant from their home, and when at home lives as an exile.

Smith (1978, 3) based her typology in contrast on the objectives of a trip, described through its content. She identified:

- *ethnic* tourism: the experience of the cultures of indigenous and often exotic peoples;
- *cultural* tourism: the experience of vanishing but familiar lifestyles;
- *historical* tourism: the experience of the glories of the past;
- *environmental* tourism: the experience of an alien environment; and
- *recreational* tourism: the 'freedom to indulge in the new morality'.

These segmentations have remained the main ones used in cultural and landscape tourism, but still await full validation of their discreteness through multivariate analysis. More recently the measured experiences of tourists at cultural and landscape attractions have begun to be used as the basis of more specific segmentations (e.g. Prentice *et al.* 1993; Otto and Ritchie 1996) offering the potential to develop attractions with specific market segments in mind. Solomon and George (1977) demonstrated an antecedent application of this more specific approach derived from tourist opinions, which segmented tourists interested in terms of their interest in history and sought to describe them in terms of lifestyles.

'Aesthetic cosmopolitanism' and insight

On the limited evidence available, even heritage 'club' members generally seem not to organize their holidays around heritage visits (McCrone *et al.* 1995). For most tourists 'consuming' culture and landscape, this would seem to be part only of their trip, incorporating as a minimum the setting for their visit. Urry (1995) has termed this style of consumption *aesthetic cosmopolitanism* (Table 13.1). However, in some contradiction, income polarization and a growing fear of crime are increasing demands for areas frequented only by 'people like us' (Henley Centre 1996). This implies for many tourists a continued mediation of their relationship with the cultures and landscapes visited.

Similarly, for many in the transience of visiting, with little subsequent reinforcement of ideas but many other subsequent experiences, the 'depth' and 'lastingness' of understanding gained from visits to specific attractions is unlikely to be great, and a blurring of experience can be expected. This implies that the experience of tourists needs to be measured both at the level of specific attractions and in terms of their whole trip, and the developmental process represented by memory studied. Due to the quality of tourist learning, a question arises as to an appropriate term to describe the learning

Table 13.1 Urry's 'aesthetic cosmopolitanism'

- Extensive travel and the assumption of the right to travel
- Curiosity about places
- Openness to other people
- Willingness to take the risk of moving outwith the tourist 'bubble'
- Location of one's own culture historically and geographically
- Aesthetic judgements made about places
- Semiotic skill in detached interpretation of mediated tourism

---

Source: Urry 1995

RICHARD PRENTICE

experience actually received by many visitors to cultural or landscape features. The word *insight* best conveys the quality of this experience: a perception into the character of a thing. This is very much a holistic conception of experience.

If the above demands, and the fact that this cultural and landscape tourism is likely to remain extensively a mediated experience, are both recognized, tourism development has to be seen as dependent not just upon meeting environmental preferences as cultural or landscape settings, but also upon facilitating insight and other consumer demands. Aesthetic cosmopolitanism may be thought of as implying the following research needs in cultural and landscape tourism:

- tourists' experiences: as presence, anticipation, memory and virtuality
- tourists' environmental preferences and their facilitation
- the effectiveness of media in facilitating insight among tourists
- other demands by consumers and their facilitation, and
- in consequence, tourists' liking of, or endearment to, place.

Research strategies need further to recognize that these aspects are likely to vary by market segments.

Cultural tourism and multinational integration

At the domestic level, 'foreign' ways and habits are increasingly being selectively incorporated into native cultures by Europeans and North Americans. This is especially true for 'foreign' cuisine as native kitchens are subject to growing internationalization (Henley Centre 1996). The process of European unification is encouraging such incorporation as the following European Union tourism objectives illustrate. Tourism is seen variously as:

- an export earner and employer
- a means of self-actualization and personal development
- a means of cultural appreciation
- a means of international understanding

In terms of the consumer of tourism the final three objectives are pertinent:

In a world where social and family ties have been loosened, culture is one way of relating openly with others.

(European Commission 1995: 2:7)

Culture is the sphere in which regional and local identities find their clearest articulation. . . . In a world in which institutionalized trade and a globalized economy are the rule, cultural levelling constitutes an impoverishment. Social groups are reacting to this by seeking out cultural products that reflect local and regional identities.

(*ibid.*, 2:6)

To help develop and disseminate local and regional identities while, to cite the Treaty [on European Union], 'bringing the common cultural heritage to the fore', and, '... Improvement of the knowledge and dissemination of the culture and history of the European peoples'.

(*ibid.*, 2:7, 2:15)

The potential of tourism to contribute to a European identity has been expressed in quite idealistic terms by the EU Commission:

When the citizen is a tourist and no longer on home ground, he [*sic*] is certainly conscious of the difference in culture between the Member States. . . . The strangeness which can be perceived in the use of a different language or in other ways of thinking, in other customs or even other interests, is not, in a tourist context, always a brake on exchanging ideas, but – on the contrary – may arouse a desire or wish to find out about these strange ideas and understand them.

(Commission of the European Communities 1995: 16)

In view of such policies, the means and extent of demand stimulation for international cultural tourism become pertinent issues in understanding consumers' demands.

CHARACTERIZATION OF CULTURAL AND LANDSCAPE 'CONSUMPTION'

The research agenda outlined as flowing from aesthetic cosmopolitanism merits integration into an operational framework. As the core product of cultural and landscape tourism is the beneficial experiences gained by consumers, these should be the objective of such a framework (Beeho and Prentice 1995). Sub-optimality is an increasing conceptualization of tourist experience in cities such as Oxford or Venice (Glasson *et al.* 1995). In the former, traffic and crowding have been found to head the dislikes of tourists, attributes mirrored in the views of residents. However, the relationship between tourists' experiences and benefits perceived has not been extensively explored. If a tourist is seeking cultural immersion, the presence of other tourists may deny this. However, for viewing historic townscapes, tourists may simply ignore other tourists by looking above or beyond them, or indeed welcome them as guides as to where to view. After all, in a theatre others are accepted in the audience as long as one's view of the performers is unobstructed. Most work on *perceptual carrying capacity* (the relationship between tourist volume and experience) has focused on outdoor recreation or beach behaviour (Glyptis 1991). The effect of crowding, on not only the experience of a city or landscape visited but also on the benefits gained or lost from that experience, is likely to increasingly concern tourism management as demand increases in volume.

It is necessary to begin with an understanding of what tourists actually experience when visiting cultural and landscape attractions (*cf.* Beeho and Prentice 1995). A parallel may be found in transportation planning. Road travel has been described as *kinaesthetic sensations* (Appleyard *et al.* 1964), an experience of space and motion felt in continuous sequence. Somewhat similarly, a visit to an art gallery may involve a flow of experiences, such as:

appreciation	boredom	fun
realization	fixation	imagination
peace	relaxation	stimulation
mystery	insight	reflection
understanding	focus	transposition
refreshment	amazement	tiredness
inspiration	uplift	escape

A challenge is to capture these experiences and to assess their pertinence to the tourist's visit to the attraction and, overall, to the place visited. Demand hierarchies exist in consumer research which attempt to do this, and provide potential models for use in cultural and landscape tourism.

One such hierarchy may be termed the *Manning-Haas-Driver-Brown Sequential Hierarchy of Recreational Demands* (Manning 1986; Prentice 1993c; Prentice and Light 1994). This hierarchy identifies four levels of demand: activities, settings, experiences and benefits. Level 1 represents demand for activities themselves. Level 2 concerns settings, including environmental, social, and management settings; and Level 3 concerns the experience of these settings for particular activities being pursued. Finally, Level 4 demands refer to ultimate benefits which come from satisfying experiences derived from participation, and can either be psychological or societal. The hierarchy is sequential as the dependency of benefits on the earlier levels is implied.

An alternative hierarchical framework is termed the *means–end chain* (Gutman 1982; Perkins and Reynolds 1988; Klenosky *et al.* 1993; van Rekom 1994). It specifically focuses on the links between the attributes that exist in products (the *means*); the *consequences* for the consumer provided by the attributes; and the personal values (the *ends*) which the consequences reinforce (Reynolds and Gutman 1988). Simply expressed, the means–end chain works on three levels of abstraction: product attributes, consequences and personal values. For the understanding of tourist experiences and benefits, a focus on the consequences of attributes (which may be attributes of activities or settings) is important. Of critical importance to operationalizing this methodology is the *why* question; for example, 'Why, or in what ways, did you find that beautiful?' or 'Why is that important to you?' This methodology is known as *laddering*. In using laddering the researcher should not assume that tourists have linear ladders, from the concrete to the abstract. Use suggests that these ladders are circular as well as linear, and in the Manning *et al.* terminology, that higher levels may be explained through lower levels. Such circularity is a reminder that these hierarchies are conceptualizations which managers impose on the experiences of others.

These two hierarchies have been combined into *ASEB Grid Analysis* (*A*ctivities, *S*ettings, *E*xperiences and *B*enefits), taking the levels of the first hierarchy and informing the higher levels using the laddering methodology (Beeho and Prentice 1995, 1997). ASEB Grid Analysis may be thought of as a focused *SWOT Analysis* (*S*trengths, *W*eaknesses, *O*pportunities and *T*hreats) (Piercy 1991). As its name implies, ASEB Grid is a tabular analysis with four columns representing the Manning *et al.* levels, and four rows representing the four SWOT elements. Extensions may include removing motivations explicitly from the Manning *et al.* levels, and entering them as an additional column, preceding the four. Such extraction recognizes that motivations may prime the other levels; ASEB Grid in effect becomes *MASEB Grid* (*M*otivations, etc.) with this modification. No sequential dependency of the ASEB levels need necessarily be assumed for the matrix to be informed. Instead, such relationships are an objective for testing using the matrix.

The agenda which flows from aesthetic cosmopolitanism not only includes the purposive sequence represented by the ASEB elements; it also includes demands and a time sequence. The demands include those of environmental preference, which at minimum is for setting, and demands for insight. The time sequence ranges from expectations, through presence, to memories; with virtuality a potential addition. These dimensions may be defined conceptually as a multiple consumption matrix (Figure 13.1) which summarizes the kinaesthetic sensations which experience provides (Prentice 1996). Separate matrices relating demands and temporal sequence can be produced for each of the ASEB levels, and potentially for the MASEB levels if motivation is separately defined. The matrix approach is useful both in applications where the ASEB sequence is found to be causal, as well as where it is not. As virtual technologies develop, a fourth dimension may in the future be added to the schema, differentiating reality and virtuality as dimensions. It is the informing of these matrices which presents a challenge for research in cultural and landscape tourism.

MEETING DEMANDS: PREFERRED ENVIRONMENTAL SETTINGS

More is known about dimensions of landscape preference than about dimensions of cultural preference in tourism. Indeed, the latter have often been assumed and summarily labelled for advertising (in terms such as 'colourful', 'quaint', 'romantic') rather than empirically determined. Literature on preferred landscapes derives from several different sources. These include environmental psychology, landscape appreciation and planning, economics and leisure science. Each of these literatures needs to be reviewed for

Definition of a Research Agenda	Environmental Preference	Insight	Other Demands	Endearment
Expectations				
Presence				
Memories				
Virtuality				

A separate matrix can be produced for activities, settings, experiences and benefits, thereby adding a third dimension:

The consumption matrix becomes, in effect, an exploded cube.

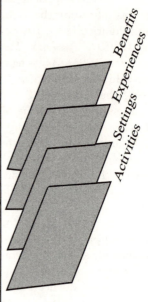

Benefits
Experiences
Settings
Activities

Figure 13.1 Heritage tourism multiple consumption matrix
Source: Prentice 1996

potential contributions to an understanding of preferred settings for engendering experiences through tourism.

Environmental psychology

Much of the relevant psychological research is limited by its generic approach to preference, having sought to measure generic visual quality rather than landscape utility. General environmental preferences have been identified, rather than preferences pertaining to particular activities or roles, such as being a tourist.

The older literature on landscape preference may be characterized as inductive in style: empirical studies, leading to theoretical generalizations about *environmental cognitive sets* and their derivation. An environmental cognitive set has been defined as 'essentially a way of perceiving or thinking about one's surroundings . . . a plan to select specific types of information for processing and/or to perform specific mental operations on information being processed' (Leff and Gordon 1979). Potentially, two process models have particular relevance to the understanding of tourists' experience of landscape: that of Ulrich, and that of Kaplan and Kaplan. These authors postulated the universality of certain preferred landscape features transcending particular landscape types.

Twenty years ago, Ulrich (1977) proposed a model emphasizing *mystery* and *legibility*. The former was defined as the 'promise of information', in the sense that additional information could be gained by entering an area and exploring it, subject to the perceived risk being low. Legibility was defined as the ease with which the landscape could be comprehended, and involved concepts of identifiability, organization, clarity, focality, ground texture, depth and complexity. Ulrich postulated that 'visual landscape preference is . . . a response in favor of scenes which efficiently transmit landscape information (legibility) and which convey a sense that additional information can be gained (mystery) at low risk' (Ulrich 1977: 281). A decade later this model was restated in more elaborate form, Ulrich (1986) postulating that liking for unspectacular 'natural' scenes would increase if:

- complexity, or the number of independently perceived elements of the scene, was moderate-to-high;
- the complexity was structured to establish a focal point, and other order or pattern, and easily grasped;
- there was a moderate-to-high level of depth of scene which was clearly defined;
- the ground surface had even or uniform textures which were relatively smooth, and the observer judged that the surface was favourable to movement;
- a deflected or curving sight line was present, conveying a sense that new

landscape information lay immediately beyond the observer's visual bounds;

- judged threat was negligible or absent;
- water or park-like elements were present; and
- any human-made elements were perceived as in harmony with the 'natural' background.

Kaplan and Kaplan's (1982) and Kaplan's (1987) alternative model emphasized that preferences for landscapes are likely to be for those most likely to meet needs, by offering the promises of *involvement* and *making sense*. Kaplan and Kaplan defined making sense as *coherence* and *legibility* (immediate and future, respectively), and involvement as *complexity* and *mystery* (immediate and future, respectively again). Kaplan (1987) argued that environmental cognition need neither be a conscious process, nor one necessarily involving calculation, but rather that aesthetics was a fundamental set of inclinations. Critically, 'the rapid assessment of what the environment holds in store is assumed to be automatic and unconscious' (*ibid.*: 24), that landscape preference-making was 'an automatic assessment of the possibilities of an environment, accompanied by an immediate affective code' (*ibid.*: 25). Such a conclusion would imply that tourists are unlikely easily to be able to identify *why* they prefer one landscape to another, and that these reasons may need to be inferred. Analysis may have to stop at description, at the identification of *what* is liked rather than why it is liked. At best, tourists' rationalizations of their preferences will, from the Kaplan view, be *post facto* and attributions may not necessarily identify the true reasons. As a minimum, in-depth approaches to the identification of preferences are implied as necessary components of research designs.

Landscape appreciation and planning

Appleton (1975) developed 'prospect-refuge' theory as a model of preferred landscapes from an analysis of landscape painting. *Prospect* he defined as the visual satisfaction of the observer, that their immediate environment is free of danger through unimpeded opportunity to see; *refuge* as the guarantee of security through the opportunity to hide.

More recent approaches have sought to highlight diversity rather than universality, emphasizing in their conceptualization of value 'subjectivity and dependence upon personal history, cultural inheritance and idealized conceptions of the world' (Jacques 1995: 91). These approaches assume that beauty is not sensed automatically, but with the intervention of the intellect. As such, this literature also contrasts with the environmental psychology of Kaplan (1987) already discussed. This literature has looked at different social and personal constructions of the environment made and used by different groups (Simmons 1993). Attention has been directed to the visual and literary arts, especially paintings and the prose of place, as the subject matter for people's understanding of the meaning of place (Daniels and Cosgrove 1993; Herbert 1995).

For the past thirty years transportation planners have addressed the issue of preferred landscapes in terms of highway design. This attention has included the 'view from the freeway' (Appleyard *et al.* 1964; Urban Advisors to the Federal Highway Administrator 1968): 'The road itself furnishes an essential thread of continuity, but it must be supported by successions of space, motion, orientation and meaning' (Appleyard *et al.* 1964: 17). The objective of design becomes one of providing a sequential and unfolding flow of images.

Economics

Much of the economic literature on landscape preferences until recently has been summary in style, as it addresses a 'price' to be placed on landscape, rather than the aspects of the landscape which are valued. The economic methodologies may be categorized into those which require some form of customer survey to elicit expressed values as *stated* preferences (for example, *contingent valuation*), those which use survey data other than stated preferences (for example, recent developments using the *travel-cost method*) and those which measure observed behaviour as a proxy for values as *revealed* preferences (for example, *hedonic pricing*).

Contingent valuation directly asks the respondent for their willingness to pay towards the preservation of an asset (Bateman *et al.* 1994; Grosclaude and Soguel 1994; Willis 1994; Tunstall and Coker 1995). Contingent valuation assumes hypothetical but structured markets to which respondents can give true valuations. It may be applied both to use-values (for example, of visiting a landscape) and to non-use values (for example, from knowing that a landscape exists although the beneficiary never actually visits it). However, the hypothetical nature of contingent valuation means that it has been found easiest to apply to things related directly to respondents' experiences; for example, a forest drive rather than the whole forest (Hanley 1989).

Contingent valuation requires that individuals express their preferences for some environmental resources, or changes in the status of resources, by answering questions about hypothetical choices. This is approached in two ways: first, by questioning about how much the respondent is willing to pay for a welfare gain, so to ensure that this occurs, or pay to prevent a loss; second, by questioning about how much the respondent is willing to accept in order to tolerate a loss, as compensation for a welfare loss, or to accept in order to forgo a gain (Bateman and Turner 1993). It needs to be recognized that if uncritically used, contingent valuation is inherently susceptible to bias. An individual may deliberately underestimate or overestimate his or her willingness to pay. However, applications to cultural and landscape tourism have included Bateman *et al.* (1994), Garrod *et al.* (1994), Grosclaude and Soguel (1994) and Willis (1994), with the methodology increasing in popularity.

The travel cost method assumes that landscape value is related to the travel cost incurred in reaching that landscape (Willis and Garrod 1991; Randall 1994). Whereas such a method may be valid for long journeys, the logic of interpreting travel costs as a valuation of the destination is less clear for more local travel which may be multifaceted. The method also only considers users, persons who have made the journey to the destination, thereby ignoring other types of consumption.

A general model for the travel cost approach is as follows (Prentice 1996):

$$V = f(T, A_1 \ldots A_n, R, I)$$
where:

V = the number of visits made by an individual to a destination;

T = the travel cost faced by the individual in visiting this destination (which may either be defined in money or time costs);

A = whether the individual is using the resource for activity $_1$ to $_n$ (often so-called nominal 'dummy' variables);

R = rating given by an individual for a particular destination;
and
I = Income of the individual concerned.

Because of the diverse range of activities undertaken by visitors, equations of this kind can become quite complex (Willis 1991; Willis and Garrod 1991), with many of the variables specified in only nominal form.

Hedonic pricing is perhaps the most simple in concept of the three methods. It attempts to evaluate environmental attributes from among other attributes determining prices actually achieved in the market place (Garrod and Willis 1992; Kask and Maani 1992; Bateman 1993). The price of an environmental good is imputed by examining the effect its presence has on a relevant market-priced good.

MEETING DEMANDS: INSIGHT

Insight is usually facilitated through the explicit *interpretation* of aspects of the destination to the visitor. That is, through identification and explanation of what may be seen or imagined at a place. This insight may be gained by using unreal as well as real things. Variation in demands for authenticity was inherent in Cohen's (1979a) typology, with his recreationalist and diversionary tourists seen as consumers of the inauthentic; players in a game with the enjoyability of the occasion being dependent upon a willingness to accept make-believe. However, only when the staged is assumed to be real is gullibility present (Cohen 1979b). Gullibility may be an appropriate test if insight is intended to be facilitated. In consequence, an over-ready acceptance of inauthentic provisions may compromise non-economic policy objectives, such as international understanding.

Interpretation can take place at different levels of objective. Most simply, it is pointing out of what may be significant in that which is being viewed. More elaborately, interpretation may enhance the experience of a place, and may prompt wonder. More elaborately still, interpretation can facilitate the understanding of the social, economic and physical environmental systems inherent in a place, through the enhancement of learning. As such, interpretation as a minimum may not only have an informational function, but also an experiential or educational function. However, other objectives are also common beyond this minimalist stance. Interpretation is often about values. It is also frequently a process of communicating to people the significance of a place so that they can enjoy it more, understand its importance, and develop a positive attitude towards conservation. As such, interpretation is used to enhance the enjoyment of place, to convey symbolic meanings, and to facilitate attitudinal or behavioural change. The latter is frequently an objective of interpretative provision in national parks. Implicit in the recognition that interpretation is value-laden is the questioning of *whose* values are being interpreted to *whom*, and *why* these values are accepted.

Interpretative media are now diverse in cultural and landscape explanation. They include people, music, smells, displays, notices and labels, diagrams, information panels, models, 'touch-tables', guidebooks, puzzles, guided walks, period rooms, audio-visual techniques, so-called 'dark rides', simulated animation (a simulated ride with the vehicle actually staying in the same place), virtual reality, other computer interactive techniques, and theatre. Modern electronics increasingly offers the potential to separate the need to visit a place from its interpretation, through virtual reality systems, multimedia, and across greater spatial distances, the Internet (Cartwright 1993; Williams and Hobson 1995; Bath 1995). Many contemporary tourist attractions facilitate the illusion of being in a different time or place. In their imagination tourists are taken back in time, as in a historical drama, novel or film, or placed in a different physical environment.

> Experience the mysterious and magical world of the Celts. . . . Where nothing is quite what it seems. . . . Experience the history and culture of the Celtic people as Nia and Gwydion guide you through a mystical world and take you on an unforgettable journey into the past and back into the future . . . a dramatisation of life in a Celtic settlement where visitors interact with the characters, a journey into the Celtic other-world and a theatrical presentation of the Celtic future.
>
> (*Celtica*, promotional leaflet, Machynlleth, 1995)

> Savour the atmosphere of a medieval ale house! . . . Imagine being

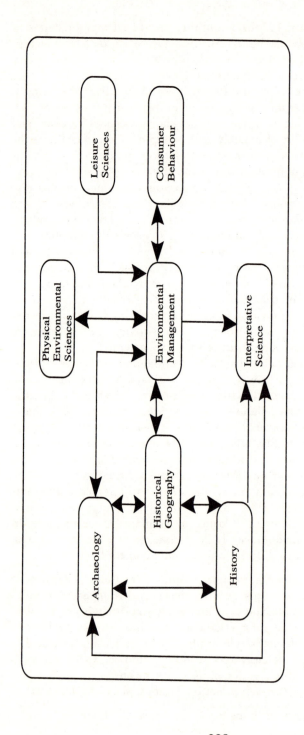

Figure 13.2 Management knowledge bases for cultural and landscape tourism and their potential interlinkage with interpretative science
Source: Prentice 1996

transported back in time to the Middle Ages and experiencing daily life in an Irish Medieval town. This is your treat at Geraldine Tralee.

(*Kerry the Kingdom*, promotional leaflet, Tralee, 1995)

Of the formal evaluation of the effectiveness of media, most has focused on what visitors notice at attractions (Herbert 1989; McManus 1989; Light 1995a); their experiences while visiting (Moscardo and Pearce 1986; Prentice *et al.* 1993; Beeho and Prentice 1995); what they learn or fail to learn during their visit (Prentice 1991, 1993c, 1995; Light 1995b; Ryan and Dewar 1995); consequential involvement in projects (Gowing and Major 1995); or on their memories of an attraction subsequent to their visit (McManus 1993). Despite the prominence of these media in delivering cultural tourism products, as yet too few studies have been undertaken to enable the making of other than very hesitant recommendations about the effectiveness of individual media, or of media mixes.

Interpretation is inherently multidisciplinary, and at best interdisciplinary (integrating different disciplinary inputs). To date interpretation has largely been taken from the outputs of other disciplines. Figure 13.2 presents a more optimistic picture, a desired future so to speak (Prentice 1996): not only is interpretative science shown as more integrated into the disciplines, it is also shown as affecting the content of other disciplines, giving it a proactive role. Figure 13.2 offers a schematic future for interpretation as product development in cultural and landscape tourism, and is elaborated in Figures 13.3 and 13.4 in terms of the types of content the inter-linkages represent.

CONCLUSIONS: THE CHALLENGES OF FACILITATING MEANING IN THE CONTEXT OF SUSTAINABILITY

This chapter has offered a largely demand-side analysis of cultural and landscape tourism in the context of sustainable development. It has sought to show how the understanding of preferred tourism environments and demands for insight may be defined, measured and appraised. Without the knowledge of what tourists are seeking in terms of experiences and benefits, the effects of management strategies are likely to be haphazard. Comparable attention needs to be given to understanding how tourists experience the cultures and landscapes they visit as to that previously given to understanding impacts: indeed, such attention is likely to be essential if impact management is to be effective and wider policy objectives of international understanding are to be achieved. As such, tourism will only be sustainable if it is simultaneously in harmony with hosts, environment, policy objectives *and* tourists' demands.

Implicit also in the approach presented here is the understanding, and particularly the segmentation, of demand for cultural and landscape tourism

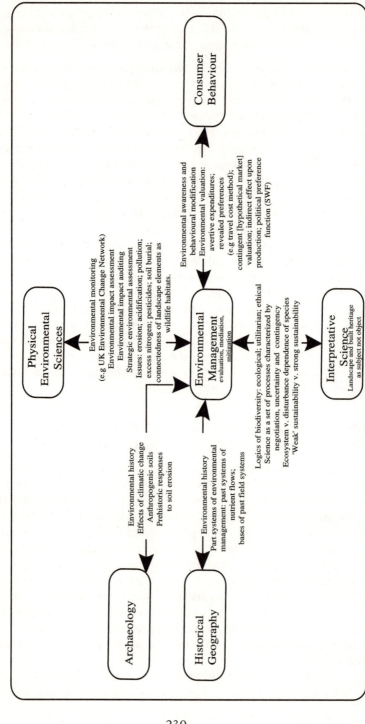

Figure 13.3 Interpretative science for cultural and landscape tourism in the context of environmental management

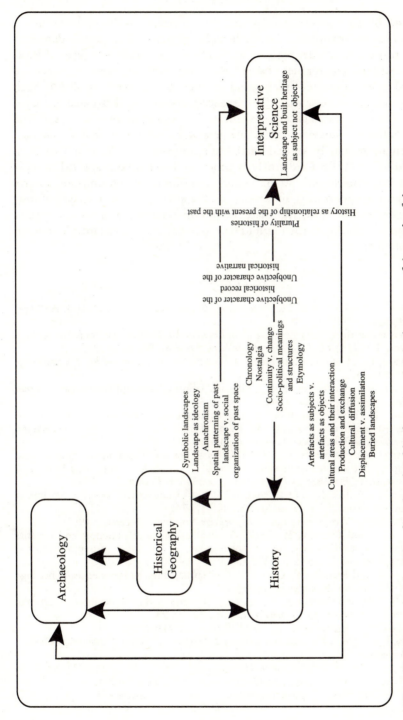

Figure 13.4 Interpretative science for cultural and landscape tourism in the context of the study of the past

Source: Prentice 1996

in terms of the resources experienced. Interpretative media provide a ready basis of such segmentation, but little work has been undertaken to date even using this basis. Broader segmentations of visitors to different types of landscape and built environment based upon experiences and benefits gained are needed in much greater number and settings than currently available. Such segmentations are essential if experience-based or benefits-based management of cultural and landscape resources is to become more widely used. The *multiple consumption matrix* offers one generic framework within which such segmentations may be set and related to motivations, activities and settings, and builds on ASEB Grid Analysis which to date has been applied to individual attractions, such as museums. The multiple consumption matrix provides the opportunity to develop such applications for destination areas as attractions. In turn, the matrices inform understanding of the tourist demand component of the fourfold harmony essential to sustainable tourism development.

REFERENCES

An Foras Forbartha (1969) *The Protection of the National Heritage*, Dublin: An Foras Forbartha.

An Foras Forbartha (1985) *The State of the Environment*, Dublin: An Foras Forbartha.

Appleton, J. (1975) *The Experience of Landscape*, London: Wiley.

Appleyard, D., Lynch, K. and Myer, J. R. (1964) *The View from the Road*, Cambridge MA: MIT Press.

Ballantine, J. L. and Eagles, P. F. J. (1994) 'Defining Canadian ecotourists', *Journal of Sustainable Tourism*, 2: 210–14.

Bateman, I. (1993) 'Valuation of the environment, methods and techniques: revealed preference methods', in R. K. Turner (ed.) *Sustainable Environmental Economics and Management*, London: Belhaven, 192–265

Bateman, I. and Turner, R. K. (1993) 'Valuation of the environment, methods and techniques: the contingent valuation method', in R. K. Turner (ed.) *Sustainable Environmental Economics and Management*, London: Belhaven, 120–91.

Bateman, I., Willis, K. and Garrod, G. (1994) 'Consistency between contingent valuation estimates', *Regional Studies*, 28: 457–74.

Bath, B. (1995) 'Virtual heritage', *Interpretation Newsletter*, December 1995: 3.

Beeho, A. J. and Prentice, R. C. (1995) 'Evaluating the experiences and benefits gained by tourists visiting a socio-industrial heritage museum', *Museum Management and Curatorship*, 14: 229–51.

Beeho, A. J. and Prentice, R. C. (1997) 'Conceptualising the experiences of heritage tourists', *Tourism Management*, 18: 75–87.

Botterill, T. D. and Crompton, J. L. (1996) 'Two case studies exploring the nature of the tourist's experience', *Journal of Leisure Research*, 28: 57–82.

Bussi, G. (1995) *Four Seasons in Cross Stitch*, London: Merehurst.

Campbell, C. (1987) *The Romantic Ethic and the Spirit of Modern Consumerism* Oxford: Blackwell.

Cartwright, W. (1993) 'Multimedia and mapping', *Cartography*, 22 (2): 25–32.

Cater, E. and Lowman, G. (eds) (1994) *Ecotourism*, Chichester: Wiley.

Cohen, E. (1979a) 'A phenomenology of tourism experiences', *Sociology*, 13: 179–201.

——(1979b) 'Rethinking the sociology of tourism', *Annals of Tourism Research*, 6: 18–35.

Commission of the European Communities (1995a) *The Role of the Union in the Field of Tourism*, COM(95)97, Brussels: Commission of the European Communities.

Commonwealth Department of Tourism (1994) *National Ecotourism Strategy*, Canberra: Commonwealth of Australia.

Convery, F. J., Flanagan, S., Keane, M. and O'Cinneide, M. (1994) *From the Bottom Up: A Tourism Strategy for the Gaeltacht*, Galway: An Sagart An Daingean.

Countryside Council for Wales (1995a) *Traditional Orchards*, Bangor: Countryside Council for Wales.

——(1995b) *European Habitats Directive*, Bangor: Countryside Council for Wales.

——(1995c) *Towards a Common Vision: Supporting Community Participation in the Local Environment*, Bangor: Countryside Council for Wales.

——(1996) *Access to the Welsh Countryside*, Bangor: Countryside Council for Wales.

Daniels, S. and Cosgrove, D. (1993) 'Spectacle and text: landscape metaphors in cultural geography', in J. Duncan and D. Ley (eds) *Place/Culture/Representation*, London: Routledge, 57–77.

European Commission (1995) *Article 10 of the ERDF. Second Programme of Inter-regional Co-operation and Innovative Actions within the Structural Funds 1995–9. Pilot Projects Relating to New Sources of Jobs. Pilot Projects Relating to Inter-Regional Co-operation for Economic Development in the Cultural Field*, Brussels: European Commission DGXVI, Regional Policy and Cohesion.

Feehan, J. and O'Donovan, G. (1996) *The Bogs of Ireland: An Introduction to the Natural, Cultural and Industrial Heritage of Irish Peatlands*, Dublin: Environmental Institute, University College.

Forsyth, P., Dwyer, L. and Clarke, H. (1995) 'Problems in the use of economic instruments to reduce adverse environmental impacts of tourism', *Tourism Economics*, 1: 265–282.

Gabriel, Y. and Lang, T. (1995) *The Unmanageable Consumer: Contemporary Consumption and its Fragmentation*, London: Sage.

Garrad, L. S. and Hayhurst, Y. M. (1988) *Samplers in the Collections of the Manx Museum*, Douglas IOM: Manx Museum and National Trust.

Garrod, G. D. and Willis, K. (1992) 'The environmental economic impact of woodland: a two-stage hedonic price model of the amenity value of forestry in Britain', *Applied Economics*, 24: 715–28.

Garrod, G. D., Willis, K. G. and Saunders, C. M. (1994) 'The benefits and costs of the Somerset Levels and Moors ESA', *Journal of Rural Studies*, 10: 131–45.

Glasson, J., Godfrey, K., Goodey, B., Absalom, H. and van der Borg, J. (1995) *Towards Visitor Impact Management*, Aldershot: Avebury.

Glyptis, S. (1991) *Countryside Recreation*, Harlow: Longman.

Gowing, G. and Major, R. E. (1995) 'The "nest test" experiment: are community involvement and good science mutually exclusive?' *Museum Management and Curatorship*, 14: 169–80.

Graburn, N. H. (1978) 'Tourism: the sacred journey', in Smith, V. (ed.) *Hosts and Guests*, Oxford: Blackwell, 17–31.

Grosclaude, P. and Soguel, N. C. (1994) 'Valuing damage to historic buildings using a contingent market', *Journal of Environmental Planning and Management*, 37: 279–88.

Gutman, J. (1982) 'A means-end chain model based on consumer categorisation processes', *Journal of Marketing*, 46: 60–72.

Hanley, N. D. (1989) 'Valuing rural recreation benefits: an empirical comparison of two approaches', *Journal of Agricultural Economics*, 40: 361–74.

Helber, L. E. (1995) *Redeveloping Mature Resorts for New Markets*, Pacific Asia Travel Association Occasional Paper no. 12, San Francisco CA: PATA.

Hendry, R. P. (1994) *The Living Model Railway*, Wadenhoe: Silver Link.

Henley Centre (1996) *Hospitality Into the 21st Century: A Vision for the Future*, Camberley: Joint Hospitality Industry Congress.

Herbert, D. T. (1989) 'Does interpretation help?' in D. T. Herbert, R. C. Prentice and C. J. Thomas (eds) *Heritage Sites: Strategies for Marketing and Development*, Aldershot: Avebury, 191–230.

——(1995) 'Heritage as literary place', in D. T. Herbert (ed.) *Heritage, Tourism and Society*, London: Mansell, 32–48.

——(1996) 'Artistic and literary places in France as tourist attractions', *Tourism Management*, 17: 77–85.

Hughes, G. (1995) 'The cultural construction of sustainable tourism', *Tourism Management*, 16: 49–59.

Jacques, D. (1995) 'The rise of cultural landscapes', *International Journal of Heritage Studies*, 1: 91–101.

Johnston, B. R. and Edwards, T. (1994) 'The commodification of mountaineering', *Annals of Tourism Research*, 21: 459–78.

Kaplan, S. (1987) 'Aesthetics, affect and cognition: environmental preference from an evolutionary perspective', *Environment and Behavior*, 19: 3–32.

Kaplan, S. and Kaplan, R. (1982) *Cognition and Environment: Functioning in an Uncertain World*, New York: Praeger.

Kask, S. and Maani, S. A. (1992) 'Uncertainty, information, and hedonic pricing', *Land Economics*, 68: 170–84.

Kelly, G. A. (1955) *The Psychology of Personal Constructs*, New York: Norton.

King, B. (1994) 'What is ethnic tourism? An Australian perspective', *Tourism Management*, 15: 173–6.

Klenosky, D. B., Gengler, C. E. and Mulvey, M. S. (1993) 'Understanding the factors influencing ski destination choice: a means-end analytical approach', *Journal of Leisure Research*, 25: 363–79.

Leff, H. L. and Gordon, L. R. (1979) 'Environmental cognitive sets: a longitudinal study', *Environment and Behavior*, 11: 291–327.

Light, D. F. (1995a) 'Visitors' use of interpretive media at heritage sites', *Leisure Studies*, 14: 132–49.

——(1995b) 'Heritage as informal education', in D. T. Herbert (ed.) *Heritage, Tourism and Society*, London: Mansell, 117–45.

Manning, R. E. (1986) *Studies in Outdoor Recreation*, Corvallis: Oregon State University Press.

Masberg, B. A. and Silverman, L. H. (1996) 'Visitor experiences at heritage sites: a phenomenological approach', *Journal of Travel Research*, Spring: 20–5.

Mason, P. (1994) 'A visitor code for the Arctic', *Tourism Management*, 15: 93–7.

Mathieson, A. and Wall, G. (1982) *Tourism: Economic, Physical and Social Impacts*, Harlow: Longman.

McCrone, D., Morris, A. and Kiely, R. (1995) *Scotland – The Brand*, Edinburgh: Edinburgh University Press.

McManus, P. M. (1989) 'What people say and how they think in a science museum', in D. Uzzell (ed.) *Heritage Interpretation 2*, London: Belhaven, 156–65.

——(1993) 'Memories as indicators of the impact of museum visits', *Museum Management and Curatorship*, 12: 367–80.

Medlik, S. (1993) *Dictionary of Travel, Tourism and Hospitality*, Oxford: Butterworth Heinemann.

Meldon, J. and Walsh, J. (1995) 'Structural funds and sustainable regional development in Ireland', in R. Byron (ed.) *Economic Futures on the North Atlantic Margin*, Aldershot: Avebury, 315–27.

Milton, K. (1990) *Our Countryside, Our Concern*, Belfast: Northern Ireland Environment Link.

Moscardo, G. and Pearce, P. L. (1986) 'Visitor centres and environmental interpretation', *Journal of Environmental Psychology*, 6: 89–108.

Ó Cléirigh, N. (1985) *Carrickmacross Lace: Irish Embroidered Net Lace*, Mountrath: Dolmen.

Orams, M. B. (1995) 'Towards a more desirable form of ecotourism', *Tourism Management*, 16: 3–8.

Otto, J. E. and Ritchie, J. R. B. (1996) 'The service experience in tourism', *Tourism Management*, 17: 165–74

Parlett, G., Fletcher, J. and Cooper, C. (1995) 'The impact of tourism on the Old Town of Edinburgh', *Tourism Management*, 16: 355–60.

Pearce, P. L. (1993) 'Fundamentals of tourist motivation', in D. G. Pearce and R. W. Butler (eds) *Tourism Research: Critiques and Challenges*, London: Routledge, 112–34.

Perkins, S. W. and Reynolds, T. J. (1988) 'The explanatory power of values in preference judgements: validation of the means–end perspective', *Advances in Consumer Research*, 15: 122–6.

Piercy, N. (1991) *Market-Led Strategic Change*, London: Thorsons.

Pigram, J. J. (1993) 'Planning for tourism in rural areas', in D. G. Pearce and R. W. Butler (eds) *Tourism Research: Critiques and Challenges*, London: Routledge, 156–74.

Prentice, R. C. (1989) 'Visitors to heritage sites', in D. T. Herbert, R. C. Prentice and C. J. Thomas (eds) *Heritage Sites: Strategies for Marketing and Development*, Aldershot: Avebury, 15–61.

Prentice, R. C. (1991) 'Measuring the educational effectiveness of on-site interpretation designed for tourists', *Area*, 23: 297–308.

——(1993a) *Change and Policy in Wales: Wales in the Era of Privatism*, Llandysul: Gomer.

——(1993) *Tourism and Heritage Attractions*, London: Routledge.

——(1993c) 'Motivations of the heritage consumer in the leisure market: an application of the Manning-Haas demand hierarchy', *Leisure Sciences*, 15: 273–90.

——(1994) 'Heritage: a key sector of the "new" tourism,' in C. P. Cooper, and A. Lockwood (eds) *Progress in Tourism, Recreation and Hospitality Management 5*, Chichester: Wiley, 309–24.

——(1995) 'Heritage as formal education', in D. T. Herbert (ed.) *Heritage, Tourism and Society*, London: Mansell, 146–69.

——(1996) 'Tourism as experience – tourists as consumers: insight and enlightenment', inaugural professorial lecture, Edinburgh: Queen Margaret College.

Prentice, R. C. and Light, D. F. (1994) 'Current issues in interpretative provision at heritage sites', in A. V. Seaton (ed.) *Tourism: The State of the Art*, Chichester: Wiley, 204–21.

Prentice, R. C., Witt, S. F. and Hamer, C. (1993) 'The experience of industrial heritage: the case of Black Gold', *Built Environment*, 19: 137–46.

Randall, A. (1994) 'A difficulty with the travel cost method', *Land Economics*, 70: 88–96.

Reynolds, T. J. and Gutman, J. (1988) 'Laddering theory, method, analysis and interpretation', *Journal of Advertising Research*, Feb–March: 11–31.

Richards, G. (1994) 'Cultural tourism in Europe', in C. P. Cooper and A. Lockwood (eds) *Progress in Tourism, Recreation and Hospitality Management 5*, Chichester: Wiley, 99–115.

Robinson, V. and McCarroll, D. (eds) (1990) *The Isle of Man: Celebrating a Sense of Place*, Liverpool: Liverpool University Press.

Ryan, C. (1995) *Researching Tourist Satisfaction: Issues, Concepts, Problems*, London: Routledge.

Ryan, C. and Dewar, K. (1995) 'Evaluating the communication process between interpreter and visitor', *Tourism Management*, 16: 295–303.

Scarman, J. (1996) *Gardening with Old Roses*, London: HarperCollins.

Simmons, I. G. (1993) *Interpreting Nature: Cultural Constructions of the Environment*, London: Routledge.

Smith, V. L. (ed.) (1978) *Hosts and Guests: The Anthropology of Tourism*, Oxford: Blackwell.

Solomon, P. J. and George, W. R. (1977) 'The Bicentennial traveler: a lifestyle analysis of the historian segment', *Journal of Travel Research*, 15 (3): 14–17.

Tourism and Environment Task Force (1996) *Tourism and the Environment Initiative Review and Future Direction*, Inverness: Tourism and the Environment.

Tunstall, S. M. and Coker, A. (1995) 'Survey-based valuation methods', in A. Coker and C. Richards (eds) *Valuing the Environment: Economic Approaches to Environmental Evaluation*, Chichester: Wiley, 104–26.

Ulrich, R. S. (1977) 'Visual landscape preference: a model and application', *Man-Environment Systems*, 7: 279–93.

——(1986) 'Human responses to vegetation and landscapes', *Landscape and Urban Planning*, 13: 29–44.

Urban Advisors to the Federal Highway Administrator (1968) *The Freeway in the City: Principles of Planning and Design*, Washington DC: US Department of Transportation.

Urry, J. (1995) *Consuming Places*, London: Routledge.

van Harssel, J. (1986) *Tourism: An Exploration*, 2nd edn, Elmsford NY: National.

van Raaij, W. F. (1993) 'Postmodern consumption', *Journal of Economic Psychology*, 14: 541–63.

van Rekom, J. (1994) 'Adding psychological value to tourism products', in J. C. Crotts and W. F. van Raaij (eds) *Economic Psychology of Travel and Tourism*, New York: Haworth, 21–36.

Williams, P. and Hobson, J. S. P. (1995) 'Virtual reality and tourism: fact or fantasy?' *Tourism Management*, 16: 423–7.

Willis, K. G. (1991) 'Recreational value of the Forestry Commission estate in Great Britain', *Scottish Journal of Political Economy*, 38: 58–75.

——(1994) 'Paying for heritage: what price for Durham Cathedral?' *Journal of Environmental Planning and Management*, 37: 267–78.

Willis, K. G. and Garrod, G. (1991) 'Valuing open access recreation on inland waterways: on-site recreation surveys and selection effects', *Regional Studies*, 25: 511–24.

14

URBAN TOURISM

Managing resources and visitors

Myriam Jansen-Verbeke

INTRODUCTION

The current success in attracting tourists to urban destinations is strongly related to a 'cultural revival' in tourism. Cultural tourism is supposed to be one of the growth sectors in the tourism market. A growing interest in cultural heritage, in historical places and as such in cities offering a diverse cultural agenda, opens new perspectives and involves new management issues for the tourism industry and for the governmental agencies. Developing tourist products based on historical and contemporary cultural resources is now a strategic option for many cities and regions in search of new economic activities. Synergy between culture and tourism sounds very promising in view of these economic targets.

In addition, this perspective finds strong support among the antagonists of mass tourism. Cultural tourism as an alternative to mass tourism, and cultural tourism as a vehicle for urban and regional revitalization, suggest both theses need to be questioned. Academic researchers, tourism marketeers and especially managers in tourism business are challenged to understand better the characteristics of this 'cultural' market segment, the push factors on the demand side and the pull factors from the supply side. In fact there still is no real consensus about the definition of cultural tourism and cultural tourists. The conservation of cultural resources and eventually the transformation process into tourism products can be a real incentive to the process of reviving cultural identity at the community or regional level. In its turn this process can create a good 'incubation' climate for development and investment in new tourism projects, which the tourism market, requiring innovation and diversification, actually needs. Where both these objectives meet, the issue of a sustainable development model becomes a high priority, in particular in cities with a limited carrying capacity.

The objective of this chapter is to contribute to the understanding of cultural tourism, the opportunities for urban tourism development and also to address the need for designing anticipatory policies and planning models to deal with the management of both resources and visitors.

MYRIAM JANSEN-VERBEKE

TRENDS IN THE TOURISM MARKET

New destinations in the international tourism market are facing the challenge to develop in a sustainable way new 'place/product combinations' which are strongly competitive, unique and attractive for different target groups. Apparently the market of 'sun, sand and sea' products has reached a stage of saturation, which explains the current interest of tourism developers – public and private – in alternative resources transformable into tourist products. In addition there is the necessity to meet the demands of a fragmented market, and this explains the drive to create innovative place products. There are signs of an important expansion of travelling for and with cultural motives, which explains the growing worldwide success and potentials of cultural tourism. Cultural tourism

- has the image of adding value to the tourist experience and hence is easily associated with 'quality tourism';
- fits into the contemporary pattern of consumption tourism, ever in search of new products and experiences which yield a high satisfaction;
- allows for wide product differentiation which is needed to meet the demands of a growing and segmented tourism market;
- opens perspectives for new destinations which cannot benefit as 'sun, sand and sea' resources;
- offers a solution for the problem of seasonality, fits in with the trend towards more active holidays, more environmentally sensitive activities, more short breaks, and added value for the business traveller.

In several European countries, the recent initiative of organizing 'open days for monuments' in combination with cultural events has proved to be a successful experiment and investment. Clearly these new products meet a changing demand pattern and at the same time stimulate a revival of interest in cultural heritage, and in sites and buildings with a history, not only among tourists but equally in the domestic market.

The fact that cultural tourism is gradually becoming an important market segment can easily be concluded from the shift in the motivation pattern of travellers. Understanding other cultures, gaining new perspectives on life and visiting cultural historical and archaeological treasures are now becoming key motives. In order to assess this trend either as a temporary and superficial shift in the market or possibly as a more fundamental socio-cultural change, it is necessary to understand more about the hidden agenda of people coming to visit a city and pretending that 'sightseeing' is their prime motive (Jansen-Verbeke and van Rekom 1996). Is this an indication of a genuine cultural interest or is eventually the combination with shopping and other leisure activities a more realistic pattern?

The key questions then are: who is this cultural tourist, what motives, activities and experiences characterize this market segment, and how relevant is this segment among the many existing tourist typologies? The market segmentation and typology of European tourists, based on an international survey of lifestyles, holds many interesting clues (Mazanec 1995). To what extent are cultural motives and activities distinguishing factors in this matrix of market segmentation? Obviously, market segments can be identified in many different ways, depending on the objectives of the study. As far as cultural interests and the activities of tourists are concerned, a simple threefold typology can be applied.

1 *The culturally motivated tourist*: This person selects a holiday destination reflecting the interest in the cultural facilities offered in the destination area. Such tourists are highly motivated to learn and to benefit from each opportunity, and they will spend several days in a particular destination (city or region) and circulate, well prepared and with a professional tour guide. This ideal cultural tourist only represents a small minority. According to a European survey in 1993 the market share would account for about 5 per cent of the so-called market of cultural tourists.

2 *The culturally inspired tourist*: Special cultural themes will attract this group; they will visit well known sites of culture, major exhibitions and festivals. They travel around and pick up experiences in many places, and never stay long in one place. With this type an element of mass tourism is evident when visiting places such as Venice, Athens, Canterbury Cathedral, etc. All want to see the same places, and mainly because of this kind of cultural tourist, visitor management policies become an important issue. According to many forecasts, a growing number of travellers will belong to this type of 'culture consumer'.

3 *The culturally attracted tourist*: The tourist, while holidaying at a coastal or mountain resort, sees an occasional visit to a city or historical site in the hinterland – a visit to a museum, church or monument – as a welcome diversion in the holiday programme. The destination is not chosen because of these facilities, but once there these opportunities may be enjoyed very much. Because more active holidays are currently getting fashionable, it is very likely that this pattern of cultural pastime will spread. Particularly for this group of tourists, cultural attractions need to be packaged, marketed as part of an arrangement and embedded in a lively urban environment. It is clear that this too is a growing market segment.

In fact this market segmentation is useful in the process of product development and marketing and raises possibilities for developing visitor management policies in a crowded or overcrowded urban destination. Despite the common characteristics, there still remains the question about the real motives of the individual, of which very little is known. Motivation is the hidden agenda behind tourist decision processes and behaviour patterns. In the search for new forms of tourism which can be appreciated as sustainable, tourist behaviour is an important an issue in the process of responsible planning.

Cultural tourism = new tourism?

According to many observers, cultural tourism can play a key role in the development of new forms of tourism and offers new opportunities for urban destinations in particular. The concept of 'new tourism' as introduced by Poon (1993) is inspired by the development principles in the Tourist Product Life Cycle, which argues that in many respects we have now reached a critical stage in the tourism growth model. This explains the present lively debate in the tourism sector on new resources, new forms of tourism and even new tourists (Jansen-Verbeke 1994). New tourism is characterized by flexibility, segmentation and diagonal integration between products, in contrast to the mass market and the standardized and rigidly packaged 'old industry'. This model can be seen as a strong 'reaction' to the uncontrolled and uncontrollable development of mass tourism during previous decades, of standardized products and of downgrading resources and tourism destinations.

However, the question as to whether this mass market is indeed splitting into diverse market segments, and what role cultural tourism may be playing in this fragmentation process, is still the subject of much speculation. So far only limited empirical research has been carried out and there is no clear evidence that cultural tourism would have less impact, positive and/or negative on the destination environment, when compared to the 'mass holiday market' (Van den Borg *et al.* 1996).

MEDIA OF CULTURAL TOURISM

There is a wide consensus about art, music and history as media of cultural tourism; but in other areas such as religion, industrial heritage, gastronomy, events and festivals, architecture, and so on, the definition of cultural tourism moves into a grey zone (Prentice 1993). In a way this also explains why cultural tourism is easily associated with urban tourism. In many countries the highest concentration of various cultural resources is indeed to be found in an urban environment (Ashworth 1995). In addition, urban tourism marketeers in particular will turn to cultural resources in order to develop tourist attractions. There are more possibilities to develop and plan for a wide and interesting tourist opportunity spectrum in an urban setting than in the countryside.

In this search for innovation and product development the emphasis has recently moved towards heritage resources which can be turned into tourist attractions (Herbert 1995). According to several authors, the development of heritage tourism and tourist products linked with history and culture offers the opportunity to introduce and implement concepts of sustainability to the planning of new forms of tourism.

SYMBIOSIS BETWEEN CULTURE AND TOURISM

It is generally assumed that culture and tourism are interdependent. The most frequently mentioned arguments supporting the view of a symbiotic relationship are based on mutual benefits. Since high costs are involved in the conservation of cultural heritage and the management of cultural facilities, tourism revenues are badly needed. On the other hand the development of culture, in its different forms and expressions, finds an incentive in tourist demand. The tourism market needs cultural resources in order to develop new products, and culture gives an added value to the tourist experience, so the interests are extremely compatible. In addition, market indications suggest that tourists are indeed interested and above all willing to pay for cultural products and experiences.

According to many observers, this relationship of interdependency has even become inevitable (Ashworth 1995). In many countries tourism has become an important sponsor of cultural heritage for which there is no alternative. From the point of view of culture, tourism is also seen as legitimation for political support, a social (and economic) justification, a means towards conservation, and an incentive for innovation. From the tourism point of view culture, in its widest definition, is seen as an ubiquitous resource which can be developed into a tourist product. To every site there is a story, and every city has its history, so the potentials appear to be unlimited. The challenge lies in the creative use of cultural resources for touristic purposes and the strategic marketing of cultural tourism products. Furthermore, the transformation process from cultural resources to tourist products is not necessarily a high-cost investment when compared to other tourist projects. Probably one of the most relevant attraction factors for investment in cultural tourism products is the fact that there is no licence on cultural resources. No ownership can be claimed and in many cases cultural resources are a public good, and a common pool where every creative tourism developer can fish (Healy 1994). Many tourism entrepreneurs in a historical town benefit from the historical environment and from the added value of the setting without substantially contributing to the conservation of this historical stage.

FROM CULTURAL RESOURCES TO TOURIST PRODUCTS: THE PRODUCTION PROCESS

In some respects the process of developing a tourist product starting from cultural resources can be compared with an industrial production process. There is a basic product, a transformation process and an end-product which is for sale on the market. This is a simplified version of the tourism production process, since in fact the final tourist product is the 'experience of the tourist' in which many different actors and factors play a role (Jansen-Verbeke 1994; Urry 1990).

This production process has two points of entry (Figure 14.1). On the one hand there are the conservation agencies in control of cultural heritage resources who hold their own values and select the artefacts and sites which in their view could benefit from tourism. On the other side there are the tourism marketeers who evaluate the cultural resources from the point of view of tourism potentials and marketing investments. The selection process can lead to different and in some cases opposing views on the strategies of product development.

The range of cultural resources which can be transformed into tourist products is indeed very wide, which explains the importance of the selection procedure. In the process of developing and planning tourism products, selecting the most interesting artefacts and sites and assigning a particular interpretation is of crucial importance and is often irreversible. Introducing cultural attractions to the mental map of the tourist is the initial and decisive step in this process of 'touristification' of cultural resources. Making history and cultural heritage accessible, transparent and attractive for tourists, implies a transformation process in which several options are open to the developers.

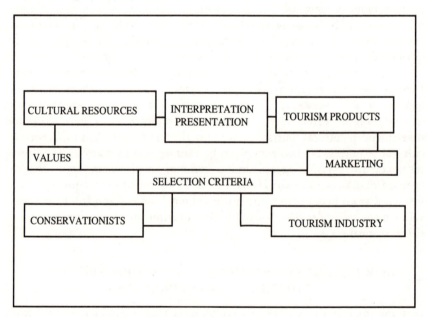

Figure 14.1 Transformation process: from cultural resources to tourism products

The accessibility of cultural and historic sites in an urban context can be developed in two different ways; in a rigorous way, the infrastructure may be adapted to the high demands of the tourism market, and open to the general public, by constructing access roads, allowing for access by coaches, creating parking facilities, minimizing transit time from the main gateways and accommodation clusters, etc. The other option is a more restricted way, with modest infrastructural modifications, possibly more difficult access, and only for the really interested, individual and public transport, which can be time consuming and perhaps less inviting. The key issue is whether infrastructural changes or improvements are a *sine qua non* of transforming cultural and historical sites into tourist attractions in a sustainable manner.

Making history and cultural heritage alive and attractive for the average tourist means developing ways in which the tourist can 'read' and 'understand' an artefact, a cultural landscape, a historic site, or a monument. In this transfer of information there are several options. First, a concentration of information at an on-site visitor centre, or a clustering of the main attractions in one place or particular urban quarter where the high standards of tourist demand can be met, is important. The concentration model has the advantage of being an effective instrument in managing the movements of tourists and hence the impact on the urban environment. This option also allows the development of new place/product combinations which are highly appreciated, such as visual presentations of the cultural heritage, sound and light shows, high-quality souvenir shops, and crafts and arts exhibitions. The prime condition however is that such a development is planned in harmony with the physical and cultural environment, in terms of scale and vernacular architecture (Soane 1994). Many tourists experience visitor centres and museums as an interesting way to 'discover' other cultures, on condition that much attention is paid to the creation of a sense of place which is, or appears to be authentic. The site which the tourist eventually 'discovers' is the result of human interaction, the way the story is communicated; it is in fact the outcome of a sophisticated marketing strategy. A minority segment of the cultural tourism market will want to move beyond this 'staged' cultural presentation and discover individually, or in small interest groups, the 'real' history and cultural heritage.

An alternative model in this process of information transfer is the development of routes and trails with a modest tourist infrastructure, which allows tourists to discover, with minor assistance (marking and signposting and/or guides), the culture and history of a city, a historical site or a region. At an early stage of tourism development in a historical city, when the destination is mainly attracting individual tourists, trendsetters, adventurers and backpackers, the latter model might be preferable, as it implies only minor environmental modifications. However, small-scale infrastructure, limited numbers of visitors and a limited impact is not by definition a guarantee for sustainable development. This is merely the first stage of a product life cycle which eventually, when successful, will evolve beyond small-scale tourism (Jansen-Verbeke 1994). As the number of tourists grows, the rules of the tourist product life cycle will soon manifest themselves. Therefore, it is of vital importance to understand the dynamics of the tourist activity and to be able to assess the impact of tourism activities in its various dimensions and in a time perspective.

TOURISM ACTIVITIES: AN INTERACTION MODEL

Clearly this new market of cultural tourism holds many opportunities for cities. Strategic planning from an economic point of view and responsible planning for sustainability need to be addressed at an early stage of the product development process. There are several constraints to this transformation process from cultural resource to tourist product. A latent conflict and eventually a manifest conflict can be anticipated by understanding the interaction patterns between tourist activities and the urban environment in which they are developed.

It is inevitable that growing numbers of tourists and an increased spatial claim for tourism-related activities in a limited space, which often is the case in historic towns, induces spatial transformation processes. Changes in the morphology of the place, in the physical structures, in the functional patterns, and in the use of public space all contribute to this transformation process. In fact, many aspects of daily life in the tourist city are affected by the tourism activities.

As a rule most studies of tourism in historic cities focus on the economic impact and the way in which the tourism activity needs to be managed and marketed in order to maximize economic benefits (Page 1995). But the impact of tourist activity is far more complex than this, and in order to plan for sustainability in the local tourism system, one needs to understand the different ways in which the tourist activity is changing the urban scene.

This assessment exercise seems an appropriate basis for developing tourism policies and as such is part of the planning process. The basic issue is that the total tourist activity intrudes in various ways in the existing interaction pattern of 'people and place' (Figure 14.2). The new activities related to tourism are interfering with the existing system and therefore hold the risk of imbalancing the system itself. The actual impact of tourism activities, of the groups of actors, and of the interaction between people (the urban community) and place (the urban setting) can be identified in three distinct dimensions: sociofacts, artefacts and mentifacts.

Sociofacts

This term refers to the way in which tourism is organized in a particular society and urban community, and the role of public and private agencies in the develop-

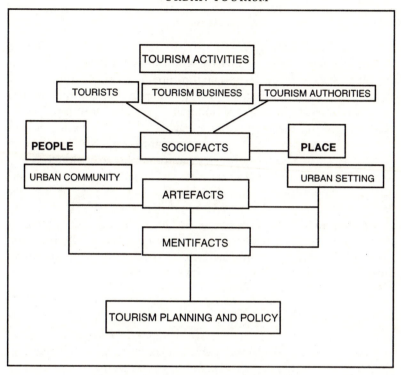

Figure 14.2 Tourism activity: an interaction model

ment of activities, including the political priorities in managing the interaction between people and place. In developing sustainable forms of tourism, the involvement and responsibilities of the different agencies concerning tourism development need to be questioned. The private sector decides on investments, on product development, on how to commercialize and market cultural heritage, and in many cases the local government stands by, watches, waits for reactions, positive or negative, before raising the issue of tourism impact on the political agenda. Sociofacts also refer to community involvement in tourism development, and the extent to which tourism is imposed – top down – by local or governmental authorities or by external agencies.

As a consequence, the economic benefits, both collective for the community as a whole and private for the local and external tourism entrepreneurs, which are induced by these tourism activities, also depend on the organizational aspects of tourism in a particular place. This can be seen as a key issue in assessing the economic carrying capacity of a tourist destination.

Artefacts

More recently, attention has been drawn to the physical impact which different tourist activities inevitably have on the artefacts in the urban environment, the changing of the morphological urban landscape through new uses, the adaptation of infrastructure for touristic purposes and the physical stress on environmental quality, on buildings, on green areas and public space. Tourists visiting historical cities are attracted by the spatial concentration of historic buildings as a setting for sightseeing and the range of opportunities for cultural activities such as visiting museums, churches, etc. (Law 1993). The number of attractions and the actual dispersion of interesting artefacts in the historic city explains to a large extent the way the tourist activity is evolving in time and space (Dietvorst 1995). Managing cultural resources and artefacts with a tourist attraction in a historic city is very closely linked with policies of physical planning (Jansen-Verbeke and van de Wiel 1995). In the current debate on the tourism carrying capacity of cities, these physical dimensions play an important role.

Mentifacts

The least-researched aspect of the interaction model concerns 'mentifacts'. This term refers to the attitudes and behaviour patterns of both guests and host community. Attitudes explain the degree of acceptance or irritation of the local community towards the intrusion of tourism-related activities affecting the quality of life and the liveliness of the place, possibly leading to conflicting values concerning, for example, behaviour in public space. The more important the tourism activity becomes in terms of tourist numbers, claims on public space and especially visibility in the daily urban scenery, the more the urban community needs to find ways of coping with the stress of tourism.

The 'social acceptance index' or 'the social irritation index' in the local community with reference to tourist activities in their daily living space can be measured in many different ways. Nevertheless it remains most difficult to assess the complex dimension of mentifacts in a quantitative way. According to previous studies, this dimension of the tourism interaction model is seen to be a relevant issue in determining the 'social carrying capacity'.

This model illustrates the complexity of the interaction patterns and hence the multidimensional character of 'carrying capacity'. It is impossible to decide on the limits to the number of tourists which a city can 'carry' in a particular time span. The impact of tourism is not only to be deduced from actual numbers of tourists, but to a large extent depends on the combined impact of tourism-related activities in a particular urban environment and the physical changes induced in the built environment.

The separate analysis of each of these relevant dimensions leads to better insight into the dynamics of interaction and eventually the threshold of conflicts. This approach may seem rather theoretical and very much in line with previous contributions on the topic of carrying capacity (Pearce 1994). It appears that progress in understanding carrying capacity and translating this into management and policy requires the input of empirical studies (van den Borg *et al.* 1996). The above interaction model was applied in the study of carrying capacity in the historical city of Bruges in Belgium, which, comparable to Venice, has become such a honeypot for cultural bees (tourists) that signs of saturation can no longer be denied.

TOURISM IN BRUGES: A CASE STUDY

The results of the case study in Bruges may be seen as relevant to many other historical cities where the pressure of tourism is building up, where questions about carrying capacity are being asked, and above all where policies for a sustainable development need to be developed. The empirical research focused on three dimensions of tourist activity in the local setting of a small-scale city.

The analysis of place characteristics included a spatial analysis of the tourist attractions, the location pattern of the core elements in the tourist product of Bruges being the historical buildings, museums, sightseeing routes and the historical urban morphology as a whole. This spatial pattern of interesting artefacts coincides to a large extent with the space use of tourists. In this process of tourist zoning, the spatial clustering of secondary product elements plays an equally important role. The wide range of facilities supporting the tourist function such as souvenir shops, restaurants, pavement cafes, hotels, parking spaces and street markets was registered and located in order to identify the physical impact of space use by the touristic infrastructure. In addition there are the tourists using, often in a demonstrative way, the public space; along the signboarded sightseeing routes, the horse-and-carriage trails, the sightseeing trip by minibus, the canal tours – all this in addition to the other traffic. This pattern of space use was also mapped in order to closely identify the key areas of tourism pressure.

The social carrying capacity, in particular the social acceptance of tourism by the local population, was investigated by means of a survey amongst opinion leaders and decision makers. The results of this survey indicated that the historic city of Bruges has indeed reached a critical phase in its development as a tourist place. Whether saturation leads to a crucial stage in urban tourism development or not, very much depends on a combination of exogenous and endogenous factors. Trends in the external tourism market are likely to affect the progress of the city along the tourist product life cycle growth model. Some examples of exogenous factors are macroeconomic trends, changing travel and mobility patterns, and not least the growing competition between urban tourism destinations in the market for short breaks and city trips.

Analysing the endogenous factors included in the 'sociofacts' implies an assessment of the views held by local authorities concerning tourism policy and the capacity (or willingness) to cooperate with the private sector in finding ways to deal with the impact of tourism. Very much depends on to what extent the tourism impact is actually considered to be a current or a future problem and therefore a political issue. The cumulative effect of daily problems such as congested traffic, shortage of parking, crowded walking areas, biased range in the retail trade, rising prices, intrusion in the private domain, and above all a general impression of overcrowding, inevitably leads to some antagonism in the local community towards tourism. Avoiding the unwanted effects of the tourism activities such as modified uses of buildings, reduction of the residential housing capacity, encroachment of hotels and tourist-oriented shops, expansion of the catering sector, rise of property values and demand for second homes by non-citizens, assumes a high priority on the political agenda.

In addition, the unequal division of costs and benefits from tourism might sharpen the antagonism. Obviously, in a community which so strongly depends on tourism, the general attitude cannot be antagonistic in every respect. Tourism is indeed appreciated as a vital resource for the local economy and even as a positive contribution to the liveliness of the place. Clearly, the degree of social acceptance is related to vested interests in tourism, but also to the daily experience of overcrowding in the urban environment. Negative statements of the local respondents were rather referring to the way conservation of historical buildings is now being handled, i.e. the policy of 'fake authenticity' as opposed to a genuine conservation policy. It is feared that the outcome of this current policy of conservation will lead to a situation in which the inner city of Bruges is looked upon as a historical stage set, an open-air museum, in which the inhabitants are assumed to play their role and become part of the tourist scene, without allowing space and forms and functions to adapt to the dynamics of modern urban life. The priorities of resource management (being conservation) and the lack of spatial management in the urban space in general were strongly criticized. This indicates that social carrying capacity is indeed determined by the physical impact of tourism as well.

The challenge of this kind of empirical research lies in identifying the factors and dynamics which eventually lead to a critical point of saturation. Indicators of a breaking point in the growth model are:

1 decline in the number of overnight visitors
2 growing absence of the original target groups
3 overcapacity in hotels leading to 'price dumping'
4 loss of quality of products and services, e.g. in the hospitality sector

248

5 decreasing interest in the city from project developers, investors and sponsors of events
6 indifferent attitudes or even an antagonistic attitude of locals towards tourism and tourists

The interesting part of these survey results is that there is indeed a limit to the intrusion and (by now) dominance of tourist space use, at least in terms of social acceptance. The fact that tourism has become part of the scenery of daily life and is affecting the quality of the urban environment and the cost of living explains the local population's strong plea for a management policy for resources with a vision encompassing management of urban space in all its aspects, not least concerning visitors. According to the local residents, policies aimed at a sustainable form of tourism should address the issues listed in Table 14.1.

Apparently the social climate in Bruges is sufficiently mature for the introduction of a tourism policy which may even include some less popular measures. Taking into account the present situation and the perception of overcrowding and saturation, especially in terms of physical capacity, it is the right time to invest in a tourism monitoring system and in visitor management policies in particular. However, there still are some serious barriers. The political climate itself, which is interested in short-term success rather than long-term sustainable tourism development, can be an impediment.

Table 14.1 Policy priorities/restrictions (% of respondents)

Hotel locations	71.5
Second homes	71.5
Traffic planning	66.2
Urban planning	60.0
Monument conservation	58.5
Residential planning	56.2
Parking planning	56.2
Pavement cafes	53.8
Location of shops	50.8
Tax regulations	38.5

Source: Survey Bruges 1990

RESPONSIBLE PLANNING

The policy of developing cultural tourism in a sustainable way implies a full understanding of the interaction patterns and the dynamics between the tourist activity and the urban setting of the cultural resources. According to Butler, failure to develop sustainable forms of tourism stems from an ignorance of tourist demands, motivations, behaviour patterns and trends, and of the impact of tourist activity (Butler 1991). So far there seems to be a lack of awareness that tourist activity indeed causes an irreversible but not necessarily a negative impact on the urban system. As shown in the Bruges case study, this impact needs to be studied in terms of organization, economic and social changes and modifications in the physical environment, including the artefacts. In many tourism destinations there is a lack of capability to implement sustainable development, or to determine the carrying capacity of the city, and as a consequence no ability to manage tourism and to develop monitoring systems, or to plan with long-term views. Probably the most serious handicap lies in the lack of understanding of the dynamics of the tourism market and the functioning of the tourist product life cycle.

In many cities, the introduction of new tourist products based on cultural resources adds a new challenge to the agenda of planning for sustainability. The promising perspective of symbiosis between culture and tourism does not yet include the long-term eroding effects of tourism. Responsible planning implies understanding of the complexity of the tourist system, the changes in demand which now tend to 'multi-motivation' and as a consequence require a 'multifunctional' supply of facilities (Pearce 1991).

The challenge is to develop or redevelop cities as destinations for cultural tourism in such a way as to assess the carrying capacity of the specific environment fully and to keep in mind the long-term impact of the tourist activity (Pillmann and Predl 1991). Looking at tourism as an integrated function of the urban system may reduce the risks of imbalance between tourism, culture and other social functions (McNulty 1993). This means defining the goals in terms of conservation of cultural resources, and finding a compromise between the short-term requests of the tourism industry with the long-term objectives of cultural heritage conservation (Jansen-Verbeke 1993). The only possible option in the interests of sustainability lies in an integrated planning approach, embracing to the cultural heritage and its environmental setting, the physical infrastructure and the tourist facilities in combination with a resource and visitor management policy (Jansen-Verbeke 1992).

RESOURCE MANAGEMENT

The presence of historical artefacts is taken for granted by many urban tourism marketeers. This common pool of cultural resources present in many cities offers plenty of opportunities to develop tourist products which meet the current market demand. However, there are constraints and rules, at least when sustainability is to be included in the agenda of tourism project developers.

Recently much attention has been paid to the issue of resource management, although the implications for cultural resources in particular are less clear than the management of natural resources. However, the number of tourism impact studies on cultural and heritage sites is rapidly expanding. In the context of historical cities, the development of strategies for integrated resource management is rather exceptional and in most cases concerns detailed studies of the physical impact of tourist uses on specific artefacts or sites. A primary and crucial decision in terms of management of resources is made when the type of touristic use is being selected. Museums, hotels, conference centres, shops or other public attractions can be strategic options, but the physical impact and the requested modifications of each are different.

In many historical places the best-fit touristic function for the old buildings is now the subject of lively debate. The opposing views, coming from conservationist forces on the one hand and tourism marketeers on the other, have been illustrated in Figure 14.1. The conservationists tend to focus on the capacity and value of the building itself, whereas the tourism developers tend to evaluate the possibilities of re-use in terms of tourist attraction and interpretation possibilities from a marketing point of view. In this selection process one could also expect some considerations concerning the qualities of the immediate urban environment, including the range of supporting facilities within walking distance. This implies an analysis of the kind of activities tourists undertake once they are on the spot and a measuring, preferably in a quantitative way, of the physical impact of the numbers of tourists having access to the artefact. The smaller the scale the more precisely the acceptable number of visitors can be decided.

At the scale of a historical town, with a diversity of places to visit and having a multifunctional role by definition, the management of particular resources can no longer be isolated from an integrated management policy. Creating alternative tourist attractions and expanding the range of supporting facilities in the core cluster can relieve the pressure in time and space on more vulnerable artefacts.

In the context of urban tourism, management of cultural resources implies the development of views on optimal uses and implementing these in the planning options and instruments (Jansen-Verbeke and van de Wiel 1995). Planning instruments which are directly applicable are the development of routes and signposted trails, the location of events, the spatial clustering of tourist facilities and the location of the tourist information and/or visitor centre. The options of clustering the tourist uses versus the dispersal model need to be balanced, taking into account both quality of the tourist experience and physical capacity of the place.

Indirectly, access to the tourist core area can be managed by strategic planning of parking areas, of traffic regulation systems and of signboards. The zoning of tourist activities also involves a location policy concerning hotels, restaurants, and especially pavement cafes, street markets and events. Obviously, these planning issues are to be integrated in an overall urban planning concept (Inskeep 1991; Murphy 1992).

A weak point in many urban tourism development plans is the fact that 'place management' is well in focus but very seldom combined with views on the actual time-space behaviour of tourists. There is a serious lack of research in this area of people's spatial behaviour patterns, concerning the constraints on the one hand and the incentives on the other. This omission cannot be accounted for by a lack of theoretical models and concepts. Ever since the emergence of geographical studies in the field of behavioural geography, launched by the School of Lund (Sweden), there have been many interesting applications published (Dietvorst 1995). Until recently the methodological implications of this kind of fieldwork discouraged many researchers, but this handicap can now easily be overcome by making use of the technical applications of Geographical Information Systems (GIS) and the adapted software. Herein lies the key to analysing and eventually understanding the spatial behaviour patterns of tourists, crucial for developing approaches to visitor management.

VISITOR MANAGEMENT

In historic cities where tourist activities determine the daily and seasonal rhythm and the quality of daily life in the urban environment, the question arises as to how to manage people, and especially visitors. Several instruments are now being experimented with to relieve the physical pressure of tourists on public space in general and on artefacts in particular. The problem of overuse and overcrowding needs to be assessed in all its aspects (Healy 1994). This statement is seen as contradictory to the interests of the tourist industry, whose motto has been 'the more tourists the more profitable the investment', which clearly reflects a short-term strategy. Supporting the concept of sustainable development means accepting that there is an upper limit, and implies the political will to invest in instruments for visitor management.

This 'new' management field is now in search of instruments and clear guidelines concerning regulations, implications, intended and unintended effects. The key issues are: regulating access to the tourist attractions (numbers and routes), reducing or deviating traffic flows, guiding the tourists, and zoning tourist use of the urban environment and its cultural resources, both in time and in space. In fact there is a wide range of instruments which can be applied to control the way visitors use urban space. The category of instruments which can be applied to regulate the different traffic flows is the most obvious one. Cars and coaches, bicycles and pedestrians:

each has its own spatial pattern which can be adapted to the objective of reducing physical and psychological pressure. In the historical city promotion of collective transport modes is particularly evident, although not always successful. Marketing collective sightseeing tours (e.g. with small electric cars or preferably by foot) might even add to the quality of the tourist experience.

More sophisticated marketing instruments include the time regulation of visitors (time tickets), the lengthening of opening hours of major attractions including price marketing, and the promotion of specific arrangements for a city visit. In fact most urban tourists are individuals enjoying the freedom of discovering a place of interest, so the promotion of guided tours can prove to be difficult unless there be an added value to this experience (e.g. visiting places which are officially closed to the public). Above all a professional guide or hostess, capable of selling an interesting interpretation of the cultural resources and well geared to the interests of the different types of tourists, can be helpful (Pizam *et al.* 1996).

In order to communicate with this group of individual tourists, the creation of visitor- and information centres proves to be most effective. Second best is communication via the accommodation sector, which of course only applies to overnight tourists and is often reduced to the business of distributing general tourist information, flyers, brochures, advertisements, to restaurants, shops, etc.

A more rigorous way of managing visitors and numbers is to choose a selective marketing strategy. Urban visitors are by definition a multi-motivated group, of which some will be more interested in the specific tourist offerings than others. Taking into account the strong and weak points of the destination, selective marketing might prove to have definite advantages. Promotion of tourist activities in the low season can be a key issue in marketing, and needs to be supported by managing the agenda of events and exhibitions in combination with price marketing (for accommodation, arrangements and package tours).

The ultimate instrument for dealing with the problem of overcrowding is the creation of new attractions, in the city or in the near vicinity, which are designed to relieve pressure on the core elements of the urban tourist product. This option raises an interesting debate on feasibility, ethics, and not least on sustainability. Knowing that the core product of many cities is strongly based on cultural resources, the question of complementary products which can be marketed to the urban and cultural visitors is now addressed in different ways. Adding 'modern attractions' in a historical setting can indeed contribute to the development of a stronger product and attract more market segments. The overall effect can be a better dispersion of visitors in time and possibly also in space, but if successful will hardly reduce the tourism stress on the urban community as a whole.

Relocation of selected tourist attractions beyond the tourist core zone can be a strategic instrument in managing the spatial behaviour pattern of visitors. However, in order to be marketable and capable of becoming part of the tourist's mental map, this needs to be a well planned concept in a carefully chosen location. A major constraint on this option amongst tourism developers in both public and private sectors is fear of competition with core elements of the tourist product. The same argument explains the reticence to market the city as part of a regional tourist product, even in cases where the hinterland might include interesting cultural resources. This could lead to a wider dispersion of visitor flows, and to some reduction of the carrying load on cities, but equally implies sharing the benefits with external partners.

Marketing strategies which could possibly and eventually lead to a reduction of the numbers of visitors do not fit into the contemporary policies of growth. There are some exceptions to this rule, e.g. historic cities and cultural sites which have indeed reached the point of saturation and are forced to manage quotas in order to guarantee quality and to consolidate their position in the market of tourism destinations.

A MISSION AND A CHALLENGE

The key issues in managing both cultural resources and urban visitors have gradually attracted the attention of different agents involved in tourism development and marketing, in urban management and in heritage conservation. The growing awareness that tourism needs to be developed according to some agreed rules and guidelines, and that anticipatory policies are necessary, reinforces the debate on responsibilities and tasks.

However, the time horizon of the different agencies varies considerably: the tourism industry focuses on seasonal and yearly business reports, local authorities are tied to an electoral cycle, and conservationists tend to look backwards, and sometimes forwards, without being much concerned (nor informed) about the dynamics of the tourism market. This difference in time horizon, and also in values, turns every tourism planning project into a process of compromises.

In addition to the possible discrepancies in short- and long-term objectives, there is the different degree of involvement of local and external agents. External investors tend to be less concerned about the long-term perspectives and impact of their tourist products on the destination environment. This could be an argument for placing the task of resource management in the hands of local organizations and authorities because they will have to live with the impact of tourism development (positive and negative), possibly beyond the stage of tourist attraction (a post-tourism stage?) and therefore are probably more likely to feel responsible for their cultural resources and their heritage. The question is whether they value this heritage in the same way that tourists do. Or is it in fact tourism which first made them aware of the value of their cultural heritage? Apparently cultural heritage is less valued if it cannot be developed and marketed as a tourist product. Changing such views might mean a long and difficult mission.

Introducing in theory and in good practice the principles of sustainability by emphasizing the irreversibility of tourism erosion tends to be seen as a negative approach and yields insufficient response. The low response to idealistic options such as sustainable tourism can be understood, because tourism planners and academics lack the expertise to translate the models and concepts into practice, and because examples of good practice are rare. In fact it is rather surprising that the issue of sustainability, which has lately inspired so many writers, has failed to appear on top of the agenda of most tourism agencies. If there is a mission for academic research in this area, it is surely to communicate the message that the future quality of tourism requires investment now, that the quality of life in an urban tourism destination needs to be carefully monitored in full understanding of the dynamics of the tourism market, and needs to be underpinned by an integrated planning vision. Cultural and heritage tourism is now at an important crossroads. The mission to market nostalgia, authenticity, education and entertainment in such a way as to enrich the tourist experience and to safeguard resources for future generations should at least match the most important current objective, which is to use tourism as a stimulus for the urban economy.

Where and how will the ideology of sustainability fit into this market-driven growth model of tourism ?

REFERENCES

Ashworth, G. J. (1995) 'Managing the cultural tourist', in G. J. Ashworth and A. Dietvorst (eds) *Tourism and Spatial Transformations: Implications for Policy and Planning*, Wallingford: CAB International, 265–84.

Butler, R. W. (1991) 'Tourism, environment and sustainable development', *Environmental Conservation*, 18 (3): 201–9.

Dietvorst, A. (1995) 'Tourist behaviour and the importance of time-space analysis', in G. J. Ashworth and A. Dietvorst (eds) *Tourism and Spatial Transformations: Implications for Policy and Planning*, Wallingford: CAB International, 163–82.

Healy, R. (1994) 'The "common pool" problem in tourism landscapes', *Annals of Tourism Research*, 21 (3): 596–611.

Herbert, D. (1995) *Heritage, Tourism and Society*, London: Mansell.

Jansen-Verbeke, M. (1992) 'Sustainable tourism development', in B. Nath *et al.* (eds) *Environmental Management*, Brussels: VUB Press, 209–27.

——(1993) 'Tourism and historical heritage management', in *European Infrastructure*, Brussels: European Community, 197–9.

——(1994) 'Tourism quo vadis? From "business as usual" to "crisis management" ', inaugural lecture, Erasmus University, Rotterdam.

Jansen-Verbeke, M. and Van de Wiel, E. (1995) 'Tourism planning in urban revitalization projects: lessons from the Amsterdam waterfront development', in G. J. Ashworth and A. Dietvorst (eds) *Tourism and Spatial Transformations: Implications for Policy and Planning*, Wallingford: CAB International, 129–45.

Jansen-Verbeke, M. and van Rekom, J. (1996) 'Scanning museum visitors: urban tourism marketing', *Annals of Tourism Research*, 23 (2): 364–75.

Inskeep, E. (1991) *Tourism Planning*, New York: van Nostrand Reinhold.

Law, C. M. (1993) *Urban Tourism: Attracting Visitors to Large Cities*, London: Mansell.

Light, D. and Prentice, R. (1994) 'Market based product development in heritage tourism', *Tourism Management*, 15 (1): 27–36.

Mazanec, J. (1994) 'Tourist behaviour and the new European life style', in W. Theobald (ed.) *Global Tourism: The Next Decade*, Oxford: Butterworth Heinemann, 199–216.

McNulty, R. (1993) 'Cultural tourism and sustainable development', *World Travel and Tourism Review*, 3: 156–62.

Murphy, P. (1992) 'Urban tourism and visitor behaviour', *American Behavioral Scientist*, 36 (2): 200–11.

Page, S. (1995) *Urban Tourism*, London: Routledge.

Pearce, P. L. (1991) 'Analysing tourist attractions', *The Journal of Tourism Studies*, 2: 46–55.

——(1994) 'Tourism-resident impacts: examples, explanations and emerging solutions', in W. Theobald (ed.) *Global Tourism: The Next Decade*, Oxford: Butterworth Heinemann, 103–23.

Phelps, A. (1994) 'Museums as tourist attractions', in A. V. Seaton *et al.* (eds) *Tourism The State of the Art*, Chichester: Wiley, 169–77.

Pillmann, W. and Predl, S. (eds) (1992) *Strategies for Reducing the Environmental Impact of Tourism*, Vienna: International Society for Environmental Protection.

Poon, A. (1993) *Tourism, Technology and Competitive Strategies*, Wallingford: CAB Publications.

Prentice, R. (1993) *Tourism and Heritage Attractions*, London: Routledge.

Soane, J. (1994) 'The renaissance of cultural vernaculism in Germany', in G. J. Ashworth and P. J. Larkham (eds) *Building a New Heritage*, London: Routledge, 159–77.

Swarbrooke, J. (1994) 'The future of the past: heritage tourism into the 21st Century', in A. V. Seaton *et al.* (eds) *Tourism: The State of the Art*, Chichester: Wiley, 222–9.

Van den Borg, J., Costa, P. and Gotti, G. (1996) 'Tourism in European heritage cities', *Annals of Tourism Research*, 23 (2): 306–21.

Urry, J. (1990) *The Tourist Gaze: Leisure and Travel in Contemporary Societies*, London: Sage.

15

THE GREEN GREEN GRASS OF HOME

Nature and nurture in rural England

Graham M. S. Dann

INTRODUCTION

In the Middle Ages, notes Urbain (1993), the countryside was considered a dangerous domain, a forbidding external territory over which travellers did not linger. Rather, they traversed this hostile space as quickly as they could, permitting themselves relaxation and enjoyment only when they reached the internal protection of the city walls. Urbain also observes that it was not until the mid-nineteenth century that the countryside came to be regarded as worthy of exploration and adventure. This period coincided with the age of Romanticism when rural areas were looked upon as ethnographic sites. Thus the poor peasant became an emblem of authenticity. The countryside was thought of as a primitive state of gentle savagery, an exotic enclave of pure nature. During the overlapping Industrial Revolution, and the subsequent decades which extended well into the twentieth century, it was the city that came to be feared. Moreover, it was against the backdrop of this urban dread that the countryside was projected as a deindustrialized, depoliticized asylum far from the madding crowds, well removed from the toxic waste and pollution associated with urbanization. The countryside was seen as a hygienic and recuperative environment, a place of genuine living and natural rhythms. In other words, pre-modernity was considered as a refuge from the excesses of rationalism and modernity.

However, it is only when we arrive at the postmodern era – the age of image – that representations of rurality in an information-based media-driven society become playful alternatives to the urban present (Joaquim 1992). Now, with the collapse of modernity, there is a search for a new consciousness, one that harks back to history and back to nature (Meyer 1990). Nostalgia thrives amid such an ethos. So too does tourism with all its ludic fantasies (Cohen 1995). Where they combine, they constitute an all-pervasive symbolic discourse of far-reaching significance. Under this strategic promotional alliance the village is now portrayed as a mythical centre where everyone knows everyone else, embryonic images of communitarian bliss are depicted in terms of roots and traditional hospitality, the hero returns to the land, and man (who previously had treated nature as a female object of conquest) becomes a friend at one with the environment (Urbain 1993). Such a reconstruction of reality, says Urry (1990: 96), constitutes 'a bucolic vision of an ordered, comforting, peaceful and, above all, deferential past'.

Furthermore, according to Lowenthal (1993: 4), we feel secure in this past: 'we are at home in it because it *is* our home – the past is where we come from'. While at first it would seem that 'we can no more slip back to the past than leap forward to the future' (Lowenthal 1993: 4), it is clear that through memory we attempt to recapture the past by recalling the lost scenes of childhood. Thus 'to be an adult means to be on good terms with the child within us, to know that we can still go home from time to time' (Lowenthal 1993: 72). Tourism, via the 'register of nostalgia' (Dann 1996) also tries to re-create 'a past out of childhood divested of responsibilities and an imagined landscape invested with all they [travellers] find missing in the modern world' (Lowenthal 1993: 25). 'If the past is a foreign country, nostalgia has made it the foreign country with the healthiest tourist trade of all' (Lowenthal 1993: 4).

Nostalgia, as Davis (1979) reminds us, literally means that painful condition associated with separation from home. However, we do not merely recall the places and events of our childhood. We filter out negative experiences and imbue the rest with special qualities. Subjective feelings about the past may therefore have little to do with actuality. Rather, the sentiment of nostalgia which they collectively constitute is a 'positively toned evocation of a lived past in the context of some negative feeling toward the present or impending circumstance' (Davis 1979: 18). This one-sided mental dialogue where the past continuously prevails, and which harbours the unexamined belief that yesterday is always better than today, undoubtedly turns on the issue of personal identity, where the question 'who am I?' is answered with reference to 'who was I?' – the simplified, prettified and otherwise distorted past of childhood where the self always emerges victorious.

Turner and Ash (1975) likewise maintain that the myth and cult of nature are promoted in the discourse of tourism whenever the latter provides images of a communal rural childhood, and that the liberation on offer becomes no more or less than a *return* to the realm of childhood. Relatedly, Boyer and Viallon (1994) indicate that, although currently monochrome photographs and poetic texts have largely replaced the romantic travellers' tales of yesteryear, these media still try to convince the tourist to go back to those locales (particularly rural settings) which evoke memories of childhood. In this connection, Squire (1994) highlights the popularity of places such as Beatrix Potter's Hill Top Farm in the English Lake District, which clearly call to mind the twin themes of countryside and childhood.

In a similar vein, Cazes (1976: 17) refers to the 'infantile regression' of the tourist, and Selwyn (1993) to the tourist as a Peter Pan who never grows up. Morin (1965) speaks of the tourist returning to the bosom of mother nature, while Jokinen and Soile (1994: 343) allude to 'the dream of the Western individual [as] a dream about the primordial symbiosis with the mother', and Dichter (1967) talks of a trip as form of birth or rebirth. Dufour (1978) goes one stage further when he treats tourism as a return to the maternal womb, where the signs of the fertile *terra genetrix* are everywhere: in the fields, the rivers and the valleys. Constrained by the myths which reside in the collective unconsciousness, tourists yearn for a return to the good old days (the myth of the Golden Age). They seek a rich, abundant and harmonious nature (the myth of the Horn of Plenty). They look for their mother in the mountains (the myth of Olympus) and in the oceans (the myth of Poseidon). Above all, however, they long for a lost childhood (the myth of the Fountain of Youth).

This essay attempts to link memories of childhood with the countryside in the nostalgic discourse of ecotourism. In fusing nature with nurture, this promotional rhetoric extends from entire nations and regions to tiny villages and hamlets. It encompasses rustic hostelries and taverns, as well as rural activities and events.[1] This presentation, however, goes beyond the position that 'phrase precedes gaze' (Dann 1996) to one which raises questions about the sustainability of a tourism that has a fixation with the past. In so doing, it extends the dimensions of sustainability, beyond such traditional considerations as resources and biophysical phenomena, to people, culture and heritage, and how the latter are variously combined to yield touristic experiences.

The setting for such an analysis is Olde England – that country which, perhaps more than any other, unashamedly basks in its former glory. Although many of its denizens today may have set aside the rural novels and poetry of Hardy and Wordsworth, and are less familiar than inhabitants of former times with the landscapes of Gainsborough and Constable, they nevertheless long to consume yesteryear through such alternative nostalgia outlets as *Past Times* and *This England*. Indeed, so deep-rooted is this yearning for the countryside that 84 per cent of the population of England and Wales visit it as tourists during the course of a year (Pigram 1993: 158). A recent UK survey[2] has also revealed that as many as 48 per cent of its respondents wanted to move out of the city to rural areas. For them, the countryside signifies an unspoilt haven, a place of order amid surrounding urban chaos.

COUNTRIES AS COUNTRY

Urbain (1993) has observed that many national tourist organizations promote their destinations as 'pays-campagne'. Ireland, for example, is portrayed as a Virgilian forgotten continent, a land of utopian calm at the end of the world, which comprises empty seashores, rugged cliffs, depopulated landscapes, winding roads, donkeys and farmsteads. Above all, it is a place of *return*. One does not go *to* Ireland; one goes *back* to Ireland. It is there that the self is rediscovered and recycled. In this green and savage land, untainted by the evils of industrial civilization and unchanged since the dawn of time, one is *reborn*. Replete with spectacles and memories of childhood, Ireland (along with Finland, Norway, Canada and a host of other green destinations which are represented as a pre-Euclidian universe) becomes 'pays enfants'. Indeed, the photographs which accompany the written publicity for such destinations are child-dominated, and their sepia tones nostalgically evoke memories of childhood. In such a manner, these territories become distinguished from 'dead countries' such as Egypt and Greece, which focus on ruins, and whose iconographies are filled with elderly peasants in traditional dress.

England, as might be expected, strives to effect a suitable compromise. It attempts to steer a middle course between hawking heritage and child-centred approaches. An advertisement emanating from the British Tourist Authority (1993) illustrates the point. The headline reads *Britain. World Capital of Tradition*. The picture shows an aristocrat sitting down to afternoon tea in the flower-filled grounds of his estate. The text identifies him as 'Piers Lord Wedgwood, the 12th generation of a family that helped define British style'. Using crockery of the same name and vintage, he is indulging in the great British 'cuppa'. The experience is said to be 'enjoyable', i.e. childlike. Since the table is set for two, the viewer is invited to fill the empty place, rather like Alice joining the Mad Hatter. Nearby on a canal there is a barge with a happy wedding couple on board. They too, as the younger set, are clearly intended to overcome the generational divide and share in the bliss. The setting is predominantly rural. Indeed, what could be more bucolic than an English country garden? All is peace and order. Even the lowly gamekeeper on the canal footpath (shades of Lady Chatterly's lover?) knows his place in this hierarchical universe of squire and peasant. 'The rich man at his table, the poor man at his gate'.

Yet this seemingly harmless publicity may not be so innocuous after all. As Lowenthal (1993: 7) points out, a pining for things past – especially things English, when 'beer was cheaper and people had more respect', and where Britain was synonymous with Empire – constitutes a longing for a time when life was simpler and worth living, a golden age of rough plenty and rude sauntering. However, says Lowenthal, this perception of the past as a flight from the present is selective, distorted and anachronistic. The reality of the situation is not so much a wistful world comprising old churches and wayside inns, but rather the bedraggled look of a decrepit village where roads were mudtracks, houses had no chimneys, 'the half timber was sagging and rotten, and the people were all bloody midgets' (Westall 1978: 156).

Thus, in spite of Kenneth Grahame's (1945: 26) rhetorical observation that 'in England we may choose from any of a dozen centuries to live in, and who would select the twentieth?' (Lowenthal 1993: 9), the truth of the matter is that having made their option, most travellers to the past would be extremely disappointed and quite unable to cope with what they encountered. However, more germane to the current presentation is Sheridan Morley's (1977: 777) warning that 'like other tourists, those drawn to the past imperil the object of their quest'. Perhaps contemplating gridlock in the Lake District and bumper-to-bumper traffic on the narrow roads of Cornwall, with patrons in hot pursuit of double cream in country cottages, Morley (1977: 777) continues: 'An eco-nostalgic crisis is on the way . . . resources will have to be preserved, revivals . . . strictly rationed' (Lowenthal 1993: 4). Alternatively stated, it is difficult to see how Merrie England can continue to be promoted *ad infinitum* without the process of tourism massification destroying the very merriness that makes it so seemingly attractive.

RURAL REGIONS

Barke and Harrop (1994) note that through promotion places are sometimes given new identities; they are presented in a different guise. Thus, for instance, South Tyneside becomes 'Catherine Cookson Country' and Swansea is transformed into 'Dylan Thomas Country'. Here quite tawdry urban centres are given a rural designation, which, when combined with a literary connection, helps alter their otherwise negative images of dereliction, decline and smokestack industries. Goodey (1994) adds that 'Shakespeare Country', 'Brontë Country' and 'Constable Country' provide the theme for rural tracts of England to be marketed as a form of 'escape culture' drawing on literary and artistic heritage, even though tourists to these areas rarely give the latter as a motivation for their visits (only 8 per cent of those patronizing Haworth, for example, profess an interest in the Brontës).

None the less, whole regions of England continue to be promoted as countries within a country. 'Captain Cook Country' (1994) is a good instance of this trend, embracing as it does an area extending from Whitby on the north-east coast in a westerly ribbon enclosed by the towns of Hartlepool and Stockton-on-Tees. The pamphlet publicizing the various sites directly or indirectly associated with the illustrious mariner dwells on the twin themes of rurality and childhood. Visitors are first placed in a situation of dependency when they are invited to 'follow in the great man's footsteps'. Thereafter they can enjoy a working farm at Newham Grange, gaze upon the schoolroom where Captain Cook was educated, eavesdrop on smugglers' tales at Saltburn's Ship Inn, or explore the Moors Centre with its adventure playground and quiz trails. However, it is the Esk Valley Scenic Line which perhaps most of all manages to combine nurture with nature, since, in the words of the promotion, 'travelling by train you can sit back and enjoy a view of idyllic rural England, untouched by the twentieth century'.

Essex, too, nostalgically appeals to the child when it describes itself as a 'county of hidden treasures'. As the copy explains, it 'is a county of small villages, ancient woodland and historic towns' (Essex Tourist Information Centre 1995). Not to be outdone is *'Lancashire's Haunted Hill Country'* (n.d.) Where visitors are welcomed to 'the land of myths, legends, superstitions and stories of ghosts and things that go bump in the night'. Apart from the well known Pendle witches, also on offer are the haunted sites of Hobstones Farm, Samlesbury Hall and Wycoller Hall. Additionally there are several ghostly hostelries going by such names as Yuticks Nest Hotel, Smackwater Jacks and Duckworth Hall Inn.

Lancashire is also promoted via its rural past. 'Step back in time' cries an advertisement for Wycoller Country Park (1995) (and by 'time' it is referring to 1000 BC!). 'Walk through ancient woodlands and past numerous waterfalls' as visitors to Ingleton (1995) have been doing since 1885, yells another. If one feels religious in *Lancashire's Hill Country* (1995) one can share George Fox's view from Pendle Hill – that vision which inspired the Quaker movement. Alternatively, if one is of a more literary frame of mind, one may wish to visit Hurstwood, where stands the cottage of Edmund Spenser, author of the Faerie Queen – or even journey to Stonyhurst, where Sir Arthur Conan Doyle studied as a schoolboy. Whatever one's present disposition, it is quickly transformed by a voyage into the past where all is in harmony with nature.

Yet just how environmentally friendly and enduring is this form of regional tourism that depends for its survival on marketing yesteryear? A recent lesson from Montana (Knize 1995) would seem to indicate extreme caution. Due to the success of Robert Redford's *A River Runs Through It*, inhabitants of that once happy rural state are beginning to speculate that the sequel may well be *A Realtor Runs Through It*. Former sleepy agricultural communities teeming with wildlife are now swamped with motels, smog and traffic-degraded natural resources, as they attempt to cater to tourists in search of paradise. Following the ancillary publicity *When Montana Sings*, depicting 'a foggy lensed idyll of children romping in wild flower meadows in the mountains' (Knize 1995: 160), the host-guest ratio is now almost 10:1. Long lines of recreational vehicles trail through the understaffed, over-capacity Glacier and Yellowstone national parks, where mountain goats lick up the drips of antifreeze. Worse still is the realization that many tourists now want to stay, and already whole areas are given over to new roads, subdivisions, wells and septic systems which have infiltrated the limited aquifers and destroyed the elk and bear habitats. At the end of this sad scenario, Knize (1995: 164) tellingly concludes: 'to Montana's promotional slogan "Unspoiled. Unforgettable" one could easily add "and Unprepared" '. Despite the hyperbole, what applies to undiscovered Montana could equally apply to the undiscovered English countryside.

HALLOWED HAMLETS

In contemporary tourism attention has been drawn to the fantasy associated with names and naming (Dann 1976). Thus travellers who reside for most of the year in towns and cities with quite unremarkable designations become attracted by the sound of places with more romantic appellations. In this regard, one relatedly comes across such delightful English country villages as Moreton-in-Marsh, Stow-on-the-Wold, Ravenstonedale, Michinhampton, Wooton-under-Edge, Whatstandwell, Deerhurst, Tottleworth, Giggleswick, Puddletown, Haltwhistle, Much Marche, Brown Willy and Nether Wallop. These places sound somehow more natural than Wigan or Watford, and a great deal more romantic than Manchester or Merseyside. They also appear to belong to the genuine past – the historic realm of the nursery – almost as if they had been constructed out of the gurgling mouths of babes.

The advertising copy associated with such rural retreats is no less fanciful. For example, there is 'the picturesque riverside town of Woodbridge ... [which] provides the ideal base to explore the tranquil villages and hamlets of undiscovered Suffolk' (Wood Hall Hotel and Country Club 1995). Or, if one prefers, one can visit Otley in Yorkshire, birthplace of Thomas Chippendale and patronized by Turner and Wesley. Otley is said to be 'steeped in history'. It has a livestock show going back to 1796 and an old-time fayre complete with carol singers and a Victorian steam organ (Leeds City Tourism n.d.)

But how viable is the appeal of rural villages as an extension of life in the city? Pigram (1993), who asks this question, and develops an answer in terms of motivational push and pull factors, indicates that this promotional process, if it is not to result in cultural conflict, but instead in beneficial economic and social interaction, has to be properly planned. Moreover, there should be a fully articulated and implemented policy that incorporates the local community into its decision making, and takes into account both the preferences and experiences of visitors. Right now, he says, research simply has not addressed these issues sufficiently. Consequently, we still need to establish an agenda which comprises the allocation of rural resources to tourism, the role of resource management agencies, and the integration of tourism policy into rural resource planning. In other words, image building, since it can result in either social destruction or construction, must be based on informed dialogue.

COSY COTTAGES AND HOMELY HOSTELRIES

Just as English villages are appealing in the very names they bear, so too are various rustic accommodations and taverns laden with this type of fantasy. As regards the former, those especially evocative of childhood can be found in such titles as 'Tyms Holm', 'Swallows Rest', 'Holly Lodge' and 'Cuckoo

Nest Farm' (*Home and Country* 1995) while those homely hostelries providing both bodily sustenance and camaraderie include such designations as 'The Stawberry Duck', 'The Malt Shovel', 'Owd Ned's Tavern' and the 'Copy Nook Hotel' (*Lancashire's Hill Country* 1995).

The descriptions of these rural establishments featured in media advertisements conjure up above all an imagery of warmth and security. In the *Guardian Weekend* (1995) for example, cottages are referred to by such epithets as 'traditional', 'welcome', 'peaceful' and 'cosy'. There is 'the cream of self-catering cottages'. There are also 'hamster cottages', 'oases of peace' and 'pretty cottages' to be found in 'the heart of England'.

One seventeenth-century inn located at West Witton in Wensley Dale (1995) promises 'daily fresh baked bread, rustic country cooking and real Yorkshire ale', while Penrhos Court (1995), of much earlier vintage (circa 1280), in Herefordshire, claims to be 'hardly touched by modern life'. More explicit is Hinton Grange (1995) near Bath in Somerset, which offers 'romance and nostalgia from an age past'. Its copy continues: 'Imagine sipping champers in a screened Victorian bath, mesmerized by your own blazing fire, then retiring to your antique four poster bed'. However, perhaps the cleverest publicity of all is the following:

> 'I say Toad, someone's converted 16th century barns near Lyme Regis into cosy cottages and called one Toad Hall. What cheek!'
>
> (Hinton Grange 1995)

In spite of their slight differences, these various promotions nevertheless share in common what Amirou (1994) refers to as a 'bulle touristique' (cf. Cohen's [1972] 'environmental bubble'). Above all, they represent havens of safety and familiarity, places of transition from the infantile dependency on the mother figure, centres of succour and schools of learning though yearning. Of related interest is the realization that many of the names given to homes in England are ones which evoke holidays of childhood (Lowenthal 1993).

However, we should not forget that there is an interactional dimension to the accommodation sector which caters to this sort of tourism. As Pearce (1990) reminds us, especially where hosts and guests are placed in close proximity, as in farm tourism for instance, each party to the encounter may have quite different goals, cognitive structures, environmental expectations, rules, roles, ways of communication and types of behaviour. Pearce's comments are equally applicable to rural hostelries and bed-and-breakfast establishments. More specifically, whereas the expectations of the proprietor are firmly rooted in the present and the future, those of the transitory visitor represent an *alternative* to the present (and future) in the quest for a cosy past. No small wonder that many agritourists display little interest in the workings of a farmyard, particularly in such distasteful phenomena as the slaughtering of animals or the castration of livestock, since activities of this genre hardly conjure up romantic images of yesteryear. In this connection,

Pearce notes that for many rural tourists, authenticity of experience resides not so much in a given place as in the people who live there. If this observation is correct, the issue of environmental sustainability assumes an interpersonal dimension, one that hinges on the balance between host privacy and visitor satisfaction.

For Buck (1977) the solution lies in erecting a front-stage of inauthenticity in order to *protect* the backstage intimacies of residents from the prying eyes of tourists. The front-stage can then be hyped as backstage *ad nauseam* without serious detriment to the locals.

TRIVIAL PURSUITS

The promotion of the great English outdoors also extends to activities and events which call forth the memories of that former idyllic state associated with a communal rural childhood. Bicycle Breaks (1995) of Colchester, for instance, advertises themed tours through gently rolling countryside, where clients can choose from 'smuggling, famous artists, steam engines, castles, natural history and real ale trails'. Alternatively, Royal Tunbridge Wells (1994), which is referred to as a country town, offers a trip back over 250 years to the eighteenth century that 'brings back to life a society filled with gaiety, politeness, fashion and flirting'. Patrons can engage in conversation with Beau Nash, join in gossiping games with inhabitants in period costume, participate in a sedan-chair race round the Pantiles, stroll along the prom, take in a Punch and Judy show, enter a competition to identify 'wotsits and thingummies', even sign on for 'language of love' workshops which provide the necessary help for learning how to write Valentine messages. As the pamphlet says, '[It's] a Georgian journey that's fun for the children, educational and entertaining for all', and if these activities are insufficient, Heritage Attractions of Great Britain (which manages the Tunbridge Wells operation) can additionally offer 'The Oxford Story', 'The Canterbury Tales' and the 'White Cliffs Experience in Dover'.

But if none of the foregoing leisure pursuits appeals, one might wish to consider Calder Valley Cruising (1995) through West Yorkshire and select from Walkley's Waterbus, World of the Honeybee Waterbus or Salmon and Strawberry Cruise. In the words of the brochure:

> Experience the rise and fall of 180 year old locks, and be legged through tunnels of time honoured fashion as you recapture the atmosphere of days gone by by our horsedrawn cruises.

Then there are celebrations, as it were, of the collective search for the 'id' (Davis 1979), such as the Old Gooseberry Fair at Egton, or the Rosa Mundi Re-enactment at Kirkleatham (*Captain Cook Country* 1994). In this connection there are also several annual festivities which are regularly featured in the 'English Diary' section of the nostalgic magazine *This England*. The summer 1994 edition, for example, contains references to:

- The Garland Ceremony at Castleton in celebration of Oak Apple Day,
- The Appleby Horse Fair – a traditional gypsy gathering dating from 1685, and
- Well Dressing and Carnival at Buxton, comprising 'Morris men, musicians, fools and customary ritual animals'.

Whereas many of these celebrations appeal to the child in the adult, some are more explicitly child-focused. Thus *Lancashire's Hill Country* (1995) for example, announces such events as a production of *Yogi Bear* and a Thomas the Tank Engine Weekend, while *This England* (1994) refers to a Duckathalon, a race involving customized baths and decoy ducks along the River Ancholme. More permanent fixtures include Windmill Animal Farm at Red Cat Lane, Burscough, and the Martin Mere Wetland Adventure which features feeding rare exotic ducks, geese, swans and flamingoes. As for child-dominated exhibitions, there is the Pendle Heritage Centre at Barrowford comprising a cruck barn with animals, woodland walks and a witches video, the Museum of Childhood at Ribchester (nine rooms filled with magic) and Britain's only cat meowseum at Ramsbottom.

More passive rural pursuits include gazing on hundreds of National Trust historic houses, or wandering around as many as 3,500 open gardens of England and Wales. The Sussex Archaeological Society (n.d.) for instance, has a programme entitled 'Treasures of Sussex Past', whereby holders of 'Remains to be Seen' season tickets gain free admission to a number of properties ranging from a Roman palace at Fishbourne to a priest house at West Heathly. Meanwhile, over in Hawes, patrons can choose from the Dales Countryside Museum – a remarkable collection of bygones, watching traditional ropemaking and seeing how the twist is put in at W. R. Outhwaite & Son, or witnessing the manufacture of hand-painted wooden toys at the Jig-a-Jog craft workshop (*Around and About Wensleydale and Swaledale* 1993). Whatever and wherever the activity, it always seems to lead back to the fond memories and forgotten era of when we were young.

But just how sustainable are these trivial pursuits, and does it really matter whether or not tourists are taken in by a series of contrived performances so long as they enjoy themselves? To answer the second question first, Bruner (1994) maintains that tourists derive satisfaction from re-enactments of the past according to the meanings they individually bestow on situations, and the latter in turn are predicated on personal biographies. In his words, 'tourists construct a past that is meaningful to them and that relates to their lives and experiences' (1994: 410). Some may come to consume nostalgia by focusing on the simple life. Others may be buying the idea of progress by contrasting the hardships of former times with the technological advances of the present. Still others may be commemorating the communal life of yesteryear and the virtues of hard work, neighbourliness and honesty – all but vanished in the context of today's rampant individualism. For these visitors, with their varying subjective filters, historical verisimilitude and the authenticity of reproduction may be less important than these features are for the custodians and interpreters of living heritage. Many recreational tourists are fully aware that they are engaging in ludic 'as if' experiences (Cohen 1995) and, in their quest for re-creation, cannot see why playing at history should not constitute a legitimate form of entertainment. Thus few of them would be deceived by Disney's Main Street USA, and many would be attracted by Colonial Williamsburg's promotion 'Spend some time in gaol. It will set you free', as they see themselves depicted grinning in eighteenth-century stocks (Lowenthal 1993: 49). Whereas 'fun for one' may be perfectly innocuous, it assumes the dimensions of an environmental problem when thousands become addicted to such publicity, and the site of yesteryear (which knew no tourists) is turned instead into a crowded playground for postmodern punters. Like its analogue in the United States, rural England can also become another theme park dispensing hyperreality, with its period-costumed inhabitants, having abandoned their traditional lifestyles, now making commoditized exhibitions of themselves for paying visitors.

THE SOCIOPOLITICS AND SUSTAINABILITY OF NOSTALGIA TOURISM

According to Williams (1974), nostalgia for rural life in England is nothing new; it has been occurring for centuries. Catering, as it does, to an essentially conservative/conservationist outlook, the rural past represents an oasis from anxiety over the present and future. It offers solid and ordered convictions about gender, identity, morality, God, law and society, in the face of contemporary relativism and doubt (Davis 1979).

What has changed, however, is that today's nostalgia, like most other facets of western 'civilization', is media generated. It 'exists of the media, by the media and for the media' (Davis 1979: 122). Postindustrial societies, such as England, come 'under a system of communication and information which constantly produces and reproduces' (Joaquim 1992: 236). Such societies seize upon pastoral images as the fanciful alternative to urbanism. They thematize the countryside. They offer it as a sanitized representation of life embodying all that is quaint, unplanned and traditional, and in the construction of such reality they *appropriate* it (Urry 1990).

Where rural space coincides with touristic space, clearly the central issue is how such space is *promoted*. Today, it is not so much a question of what attributes a destination possesses, but rather the discursive process of signification by which it becomes touristic (De Weerdt 1990). More precisely, it is a matter of who has the *power* to represent the countryside, as the battle lines become asymmetrically drawn between the urban media and local rustics

(Joaquim 1992). As an exercise in selfishness, ecotourism becomes 'ego-tourism', where city dwellers, having destroyed their own environment, seek to do the same with that of the rural Other (Wheeller 1994).[3]

For the electronic and print media, it is of little concern whether the bucolic images they project bear any relationship to reality (since otherwise they would have to represent the agricultural sector as one of the most rapidly changing domains in English society [Urry 1990]). What is more important is that such communication is targeted at a privileged audience – members of the predominantly white, car-owning service class – who wish to repudiate a modern rurality of tractors, telegraph wires, motorways and nuclear power stations, and exchange it for a world of quaint architecture and labyrinthine road systems devoid of planning and social regimentation (Urry 1990). These are the people who have come to regard their second homes in the Cotswolds as a reward for engaging in post-capitalist consumption (Craig-Fees 1996). They are the newcomers who wish to preserve all that is traditional in the heart of England. They are the ones who desire to retain the warmth and safety of their rural retreats, and in so doing maintain the myths of social harmony, continuity, stability and order. For them, the unspoilt rural represents all that is quintessentially English, a charming pastoral refuge surrounded by craft centres. By contrast, the long-term residents of these villages, who for years have been promised socioeconomic development by city-based politicians, have to forgo such claims on modernity once they and their aspirations become buried under an avalanche of collective urban nostalgia.

Gruffudd (1994) perhaps more than any other contemporary commentator, draws attention to the touristic promotion of the countryside as a form of 'ruralist cultural discourse'. He discovers its origin in several travel books published during the Second World War. This was a time, he says, when thousands of Londoners and denizens of other large English cities were being relocated to the (safer) countryside. Yet, since many of these urbanites were completely unprepared for rural life, a number of these books began to assume the characteristics of didactic texts in order to socialize their readers into a rustic world view. They were thus instructed through this 'home front' literature in how to see nature, how to gaze upon landscapes, even how to understand the language and behaviour of local yokels patronizing the village pub. In other words, the evacuees (especially those who were adults) were being placed in a childlike state. They were being treated as touristic children.

However, it is argued here that, once the guidebook has surrendered to television, and modernism has yielded to postmodernity, merely altering the medium does not necessarily imply a change of message. As a matter of fact, the same strategy of reduction to infancy seems to be employed. Thus the gardens of Castle Howard (1995) are advertised not merely as an example of outstanding natural beauty, but as the film location for television's

Brideshead Revisited, and the 'pretty villages set amongst rolling farmland' which surround King's Lynn are quickly linked to the BBC production of *Martin Chuzzlewit* (Hunstanton Tourist Information Centre 1995).

Likewise, Bekonscot (1995) in Buckinghamshire is patently reconstructed as a postmodern emblem of the past when it is promoted as the 'oldest model village in the world.' Here potential clients are invited to savour

a miniature world where time has stood still for over 65 years depicting 'Rural England' in the 1930s.

The copy continues:

Lose yourself in a wonderland of make-believe set in beautiful gardens.

Other postmodern appeals are even more forceful. Sunvil Holidays (1995), for example, urges its customers to 'enjoy life in the slow lane. To get off the tourist trail and on to the donkey track, call our brochure line now'.

It is this last piece of publicity which appeals to the anti-tourist in the tourist, or, as Urry (1990) would say, acknowledges the advent of post-tourism. In so doing, it somehow responds to the paradox highlighted by Gruffudd (1994: 261) that:

It is ironic that a tourism in part based on the promotion of rural spiritual values should be implicated in the decline of these values.

What these authors seem to be saying is that appealing to the 'green green grass of home' is not an act of reference to *our* home, but to *my* home. Publicity, even that surrounding mass tourism, above all speaks to the individual and the instant gratification of egotistical desire. Such promotional rhetoric knows, however, that in targeting like-minded persons it will be addressing millions of people, and that the outcome of this collective response will almost certainly constitute a negative environmental impact. Thus it is the discourse itself which violates through the creation of anticipation and longing. Before nature is physically exploited it is first degraded by narrative (Harkin 1995).

This essay therefore argues that it is not so much that there is currently an exodus from the city to the countryside, or the fact that over half of Britain's new homes are constructed in rural areas, or even that there are over 800 museums and 'pretend farms' exploiting the nostalgic potential of the countryside for touristic purposes (Urry 1990), but rather that such growth is unsustainable to the extent that the rural discourse of the media has replaced reality with fantasy.

As Swarbrooke (1994: 227) so succinctly points out: 'It is clear that the mass media in all its forms is now a major determinant of tourist behaviour'.

By hyping all that is traditional and childlike for post-modern consumers, it is peddling cosy and selective images in preference to truth.

Does this mean that *all* nostalgic-laden images of the countryside inevitably lead to touristic activity that destroys the very locales they seek to promote? If we accept Wheeller's (1994) view, the answer must be in the affirmative. Whether the destination be an entire nation, a region, a village or a country inn, together with the trivial pursuits carried out therein, the place and its associated behaviour are essentially the projections and outcome of hedonistic, materialistic and egotistic desire – and hence unsustainable.

However, life does not necessarily follow such deterministic patterns. When we invite children into our home, we do not expect that all the china and furniture will be destroyed. Similarly, when the tourist as child is lured into returning to the home of the countryside, degradation of rurality does not automatically ensue. There are parents and other custodians whose job it is to defend this property from violation. Furthermore, they can protect their inner sanctum by constructing a boundary of playful attractions around it, and, by charging entry fees for viewing this frontstage, can, with the generated revenue, help preserve the backstage from environmental assault. But in order for such a scenario to take effect we have to ensure that someone is guarding the guards. Can we leave that immense task to the tourism industry alone, or do we require additional input, dialogue and planning from both the visitors and the visited? This presentation suggests that the image making associated with sustainable tourism is a joint responsibility.

ACKNOWLEDGEMENT

Gratitude is expressed to John Pigram for his many useful suggestions in preparing and revising this account.

NOTES

1 In so doing, it approximates Pigram's (1993: 159) definition of rural tourism which 'spans the spectrum from low impact, seasonal, small country town festivals to year-round destination areas, featuring scenic resources and developed attractions'.
2 Data from the Henley Centre for Forecasting. The same study indicates that for the past two decades on average 58,000 people each year leave London for the countryside (Young 1995: 108).
3 For Wheeller (1994: 648), destruction of the environment is a natural human process predicated on a western growth mentality. In his words, 'There does not seem anything inconsistent in destroying our environment because to many "our environment" is, or least is perceived to be, synonymous with someone else's environment'. Stemming from this position is the whole question of whether 'doing what comes naturally' is deliberate, or an unintended side effect of materialistic selfishness.

REFERENCES

Amirou, R. (1994) 'Le tourisme comme objet transitionnel', in J. Jardel (ed.) *Le Tourisme International entre Tradition et Modernité*, Nice: Laboratoire d'Ethnologie, Université de Nice, 389–400.

Around and About Wensleydale and Swaledale (1993) pamphlet, Hawes: W. R. Outhwaite & Son.

Barke, M. and Harrop, K. (1994) 'Selling the industrial town: identity, image and illusion', in J. Gold and S. Ward (eds) *Place Promotion: The Use of Publicity and Marketing to Sell Towns and Regions*, Chichester: Wiley, 93–114.

Bekonscot (1995) advertisement in *Home and Country*, April, 29.

Bicycle Breaks (1995) advertisement in *Guardian Weekend*, 28 January, 17.

Boyer, M. and Viallon, P. (1994) *La Communication Touristique*, Paris: Presses Universitaires de France.

British Tourist Authority (1993) advertisement in *Gourmet*, May, 10–11.

Bruner, E. (1994) 'Abraham Lincoln as authentic reproduction: a critique of postmodernism', *American Anthropologist*, 96 (2): 397–415.

Buck, R. (1977) 'The ubiquitous tourist brochure: explorations in its intended and unintended use', *Annals of Tourism Research*, 4: 195–207.

Calder Valley Cruising (1995) pamphlet, Hebden Bridge: Calder Valley Cruising.

Captain Cook Country (1994) pamphlet, Middlesborough: Captain Cook Tourism Association.

Castle Howard (1995) advertisement in *Home and Country*, April, 27.

Cazes, G. (1976) 'Le Tiers-Monde Vu par les Publicités Touristiques: Une Image Mystifiante', *Cahiers du Tourisme*, série C, no. 33.

Cohen, E. (1972) 'Toward a sociology of international tourism', *Social Research*, 39: 164–82.

——(1995) 'Contemporary tourism, trends and challenges: sustainable authenticity or contrived post-modernity?' in R. Butler and D. Pearce (eds) *Change in Tourism: People, Places, Processes*, London: Routledge, 12–29.

Craig-Fees (1996) chapter in T. Selwyn (ed.) *The Tourist Image: Myths and Myth Making in Tourism*, Chichester: Wiley.

Dann, G. (1976) 'The holiday was simply fantastic', *Revue de Tourisme*, 3: 19–23.

——(1996) *The Language of Tourism: A Sociolinguistic Perspective*, Wallingford: CAB International.

Davis, F. (1979) *Yearning for Yesterday: A Sociology of Nostalgia*, New York: Free Press.

de Weerdt, J. (1990) 'Le tourisme en espace rural Français: vocation ou processus?', paper presented to the Working Group on International Tourism of the International Sociological Association, Madrid, July.

Dichter, E. (1967) 'What motivates people to travel?', address to the Department of Tourism of the Government of India, Kashmir, October.

Dufour, R. (1978) 'Des mythes du loisir/tourisme weekend: aliénation ou libération?' *Cahiers du Tourisme*, série C, no. 47.

Essex Tourist Information Centre (1995) advertisement in *Home and Country*, April, 8.

Goodey, B. (1994) 'Art-full places: public art to sell public spaces', in J. Gold and S. Ward (eds) *Place Promotion: The Use of Publicity and Marketing to Sell Towns and Regions*, Chichester: Wiley, 153–79.

Grahame, E. (1945) *First Whisper of the Wind in the Willows by Kenneth Grahame*, Philadelphia PA: Lippincott.

Gruffudd, P. (1994) 'Selling the countryside: representations of rural Britain', in J. Gold and S. Ward (eds) *Place Promotion: The Use of Publicity and Marketing to Sell Towns and Regions*, Chichester: Wiley, 247–63.

Guardian Weekend (1995) UK travel: classified section, 28 January, 18.

Harkin, M. (1995) 'Modernist anthropology and tourism of the authentic', *Annals of Tourism Research*, 22: 650–70.

Hinton Grange Hotel (1995) advertisement in *Guardian Weekend*, 4 February, 44.

Home and Country: Journal of the National Federation of Women's Institutes (1995) April.

Hunstanton Tourist Information Centre (1995) 'West Norfolk for great holidays', advertisement in *Home and Country*, April, 7.

Ingleton (1995) 'Waterfalls walk', advertisement in *Lancashire's Hill Country*, 15.

Joaquim, E. (1992) 'Différent et authentique: espace rural, espace touristique', in P. Guidani and A. Savarelli (eds) 'Gruppi e Strutture Intermedie Locali. Per una Reimmaginazione del Sistema Turistico', *Sociologia Urbana e Rurale*, XIII, 38: 235–44.

Jokinen, E. and Soile, V. (1994) 'The body in tourism: touring contemporary research in tourism', in J. Jardel (ed.) *Le Tourisme International entre Tradition et Modernité*, Nice: Laboratoire d'Ethnologie, Université de Nice, 341–62.

Knize, P. (1995) 'Montana: the perils of success', *Condé Nast Traveler*, November, 158–74.

Lancashire's Haunted Hill Country (n.d.) pamphlet, Red Rose Tourism Consortium.

Lancashire's Hill Country (1995) free guide, Blackburn: Reed Northern Newspapers (LET/Citizen) Ltd.

Leeds City Tourism (n.d.) *Otley in Lower Wharfedale*, pamphlet.

Lowenthal, D. (1993) *The Past is a Foreign Country*, Cambridge: Cambridge University Press.

Meyer, M. (1990) 'Tourism and consciousness: the Thai-farang connection', paper presented to the Working Group on International Tourism of the International Sociological Association, Madrid, July.

Morin, E. (1965) 'Vivent les vacances', in *Pour une Politique de l'Homme*, Paris: Le Seuil.

Morley, S. (1972) 'There's no business like old business', *Punch*, 29 November, 777.

Pearce, P. (1990) 'Farm tourism in New Zealand: a social situation analysis', *Annals of Tourism Research*, 17: 337–52.

Penrhos Court Hotel (1995) advertisement in *Guardian Weekend*, 22 July, 46.

Pigram, J. (1993) 'Planning for tourism in rural areas: bridging the policy implementation gap', in D. Pearce and R. Butler (eds) *Tourism Research: Critiques and Challenges*, London: Routledge, 156–74.

Royal Tunbridge Wells (1994) *A Day at the Wells: A Georgian Journey*, pamphlet, Tunbridge Wells Borough Council and Heritage Attractions of Great Britain.

Selwyn, T. (1993) 'Peter Pan in South East Asia: views from the brochures', in M. Hitchcock, V. King and M. Parnwell (eds) *Tourism in South-East Asia*, London: Routledge, 117–37.

Squire, S. (1994) 'The cultural values of literary tourism', *Annals of Tourism Research*, 21: 103–20.

Sunvil Holidays (1995) advertisement in *Guardian Weekend*, 28 January, 6.

Sussex Archeological Society (n.d.) *Treasures of Sussex Past*, pamphlet, Lewes: Sussex Past.

Swarbrooke, J. (1994) 'The future of the past: heritage tourism into the 21st century', in A. Seaton *et al.* (eds) *Tourism: The State of the Art*, Chichester: Wiley, 222–9.

This England (1994) summer edition, Cheltenham: This England.

Toad Hall (1995) advertisement in *Independent on Sunday*, 11 June, 66.

Turner, L. and Ash, J. (1975) *The Golden Hordes: International Tourism and the Pleasure Periphery*, London: Constable.

Urbain, J. (1993) *L'Idiot du Voyage: Histoires de Touristes*, Paris: Editions Payot et Rivages.

Urry, J. (1990) *The Tourist Gaze: Leisure and Travel in Contemporary Societies*, London: Sage.

Westall, R. (1978) *Devil on the Road*, London: Macmillan.

West Witton, Wensley Dale (1995) advertisement in *Sunday Times Travel*, 16 July, 5–12.

Wheeller, B. (1994) 'Egotourism, sustainable tourism and the environment: a symbiotic, symbolic or shambolic relationship?', in A. Seaton *et al.* (eds) *Tourism: The State of the Art* Chichester: Wiley, 647–54.

Williams, R. (1974) *The Country and the City*, New York: Oxford University Press.

Wood Hall Hotel and Country Club (1995) advertisement in *Guardian Weekend*, 28 January, 17.

Wycoller Country Park (1995) advertisement in *Lancashire's Hill Country*, 7.

Young, L. (1995) 'My rural affair', *She*, August, 104–8.

Part V

PERSPECTIVES ON SUSTAINABLE TOURISM

16

TOURISM AND SUSTAINABILITY

Policy considerations

Salah Wahab and John J. Pigram

AN OVERVIEW

In almost five decades since the resurgence of organized travel in 1948, tourism, as an international phenomenon, has passed through different stages until it has become a gigantic instrument in the developmental strategy of a good number of countries. As expressed by the World Tourism Organisation, tourism is ambivalent. It can contribute positively to socioeconomic and cultural achievements, while at the same time it may cause the degradation of environment and the loss of local identity. It should therefore be approached with a global methodology (Lanzarote 1995). It is incumbent upon the destination to orient its tourism growth towards meeting its socioeconomic objectives and environmental requirements. Moreover, tourism growth must coincide with the destination's prevailing value system and cultural integrity, and satisfy the needs of its local population.

This is what sustainability is all about.

The term sustainability, which has gained prominence in tourism jargon, is seen by many writers as an important part of the philosophy permeating all levels of tourism policy issues and practice from national to local (Edgell 1993). Others understand sustainability as a deep-rooted concept that relates to the fundamentals of life which sometimes are obscured by the ongoing public/private debate, regulation and rationalized government intervention.

Needless to say sustainable development and management of tourism resources should coincide with economic, sociocultural, health, safety and environmental objectives at national, regional and local levels.

The United Nations Environmental Programme (UNEP) adopted a definition of sustainable development as follows:

> Sustainable development is improving the quality of human life while living within the carrying capacity of supporting ecosystems. . . . If an activity is sustainable, for all practical purposes it can continue forever.
>
> (WTO 1995: 30)

The World Tourism Organisation (WTO) defined sustainable tourism development as that

> which meets the needs of present tourists and host regions while protecting and enhancing opportunity for the future. It is envisaged as leading to the management of all resources in such a way that economic, social and aesthetic needs can be fulfilled while maintaining cultural integrity, essential ecological processes, biological diversity and life support systems.
>
> (WTO 1995: 30)

TOURISM AND SUSTAINABILITY

It has been argued by some authors that tourism should lend itself to the concept of sustainable development because, in so many cases, tourism growth is dependent upon maintenance of the natural environment and natural processes for its own survival.

In spite of tourism emerging as an important developmental sector in the past two decades, it was considered as a 'soft option which can be developed relatively easily and does not require much in terms of specific planning or resources' (Wall 1991). But this finding is arguable and would require substantiation, as tourism has changed over the years (Butler 1993). *Tourism is no longer a soft option*. On the contrary, due to multiple variables intervening, it is becoming a rather complicated developmental sector which requires expertise and professionalism. Tourism's multifaceted function, as a socioeconomic and politico-cultural phenomenon, as a complex industry, and as a profession having its own rules and code of ethics, requires a broad intellectual background and specialized education and/or training to enable its policy makers and professionals to keep abreast of the scientific and technological changes. If we just think of the various organizational, marketing, planning and technical aspects of tourism, we must realize how complex sound tourism development is (Wahab 1975).

The old classical approach to tourism problem-solving through day-to-day tactics can no longer hold good in the face of a rapidly growing and swiftly changing world tourism industry (Wahab 1975).

While tourism has become increasingly appreciated as a contributor to the national economies of some leading tourist destinations, lack of sufficient information about the scope and essence of that activity as well as some of the processes linked to it are still widespread (Butler 1993). It is important, however, to ask what alternative forms of tourism prove to be sustainable and which do not. Some forms of tourism are considered by some writers as sustainable, such as ecotourism and special interest tourism, although as yet no empirical research has proved such claims (Butler 1993).

The crucial question in this connection is whether tourism sustainability, at least within the narrow confines of some alternative tourism forms, could be a reality.

The answer to the above question is positive. Yes, tourism sustainability could be a reality. It is the byproduct of a multitude of factors that contribute to the successful present integration and future continuity of tourism at the macro- and micro levels in destinations. As all socioeconomic, cultural, political and environmental factors are subject to change in time and space, sustainability is therefore a relative term and not an absolute fact. Markets, fashions, tastes, motivations, images, trends and destinations are all likely to change, at least in the long term. Thus tourism has to be humane and adaptive to the needs of the tourists, responsive to the needs of local communities, socioeconomically and culturally well planned, and environmentally sound. The difficulty is to determine how tourism can protect and enhance opportunities for future generations of tourists and host communities while change continues to prevail. This is the challenge that tourism sustainability has to face and surmount. Such challenge manifests itself in evolving tourist behavioural patterns and forms, progress in host community attitudes towards tourism, and change in technology and marketing opportunities. Tourism must offer products that are operated in harmony with the local environment, community attitudes and cultures, so that these become the permanent beneficiaries and not the victims of tourism development. Sustainability here does not necessarily mean that tourism maintains its viability in an area for an indefinite length of time (Butler 1993), but that it continues to be viable for future generations for a reasonably long time (fifty years, for example).

GROWTH AND DEVELOPMENT: UNITY OR DUALITY ON THE WAY TO ACHIEVING SUSTAINABLE TOURISM

The term economic development is sometimes used interchangeably with such terms as economic growth, economic progress and secular change. However, certain economists, led by Schumpeter, have made a distinction between economic development and economic growth. In their opinion, economic development is a discontinuous and spontaneous change in the stationary state which alters and displaces the equilibrium state previously prevailing, while growth is a gradual and steady change coming about by a general increase in the rate of savings and population (Schumpeter 1942). According to another economist, development requires some sort of direction, regulation and guidance to generate the forces of expansion and maintain them. This is true of most of the developing countries, whereas spontaneous growth characterizes developed countries or advanced free-enterprise economies (Jhingan 1978: 4). The simplest distinction is made by Maddison in these words: 'The raising of income levels is generally called economic growth in rich countries, and in poor countries it is called economic development' (Jhingan 1978: 4). Everyman's *Dictionary of Economics*, while using the two terms as synonyms, considers that

> economic development is used to describe not quantitative measures of a growing economy, but the economic, social or other changes that lead to growth. Growth is then measurable and objective; it describes expansion in the labour force, in capital, in the volume of trade and consumption. And economic development can be used to describe the underlying determinants of economic growth, such as changes in techniques of production, social attitudes and institutions. Such changes may produce economic growth.
>
> (1969: 136–7)

We share the opinion of those economists who see a difference between growth and development.

We define *growth* as a transformation process that aims at causing structural changes in the national economy resulting in the organized improvement of the standard of living and quality of life for the population. Such growth may happen in a developed or developing country because growth is an endless road representing people's ambition to achieve progress in every aspect of life (Wahab 1988).

Development, on the contrary, expresses the programmes that aim at the realization of a continuous and balanced increase in economic resources and the rationalization of productivity. It is a complex process that is based on scientific and applied endeavours to reach an optimum utilization of the various factors of production through technological progress and manpower development (Wahab 1988).

In tourism, as in the economic sector, such distinction is tenable. Growth presupposes the existence of a system that functions on its own within a prescribed framework. Acceleration of growth may need the intervention of public authorities and private enterprise, thus becoming assimilated with development.

Development requires the interference of the human will to plan, direct, guide and regulate progress through the utilization of available and attainable resources. If such utilization is well manipulated economically, socially, culturally and environmentally, sustainability can be realized.

The two terms, however different conceptually, would lead to the same practical conclusion, which is the fulfilment of tourist expansion, albeit through varying ways.

GENERAL CONCEPTS IN PLANNING SUSTAINABLE TOURISM

Real progress toward sustainable tourism growth has to be made through the integration of the viewpoints of at least three disciplines (Serageldin 1993):

1 That of the *economists* whose methods seek to maximize human welfare within the constraints of existing capital stock and technologies. Economists are currently relearning the importance of natural capital.
2 That of the *ecologists* who stress preserving the integrity of ecological subsystems viewed as critical for the overall stability of the global ecosystem. Environment is part and parcel of this global ecosystem.
3 That of the *sociologists* who emphasize that the key actors are human beings whose pattern of social organization is crucial for devising viable solutions to achieving sustainable development.

For sustainable tourism growth, the viewpoints of the three disciplines have to be integrated under the guidance of tourist expertise to forecast future demand, plan effective typologies (forms) and induce values for the harmonious integration of tourism development projects into a framework that is sustainable. These considerations have to be added to design criteria and land-resource use conceptualization introduced by the architect-planner.

Carrying capacity as a focal point

Tourism can be one of the effective tools for building a prosperous community economically, socially and culturally. A *sine qua non* of this is that tourism must be environmentally sound and based on sustaining the natural and cultural base rather than eroding the community's resource capital. However, tourism is not necessarily beneficial or feasible for every country or territory equally. Each destination should examine whether it has adequate attractions and facilities for tourism, and if there are potential tourist-generating markets that can endorse and be responsive to those attractions and facilities.

Carrying capacity is a central principle in environmental protection and sustainable tourism development. It determines the maximum use of any place without causing negative effects on the resources, on the community, economy and culture, or reducing visitor satisfaction.

The principle of tourism carrying capacity implies a limitation on tourism growth that may degrade scarce resources, and at the same time offers a criterion of sustainable tourism development. Carrying capacity, therefore, represents the point beyond which the tourism industry in any destination becomes unsustainable. In other words carrying capacity, whether national, regional or local, denotes how much tourism is sufficient to yield positive returns and avoid its blights.

This level is by no means easy to ascertain, especially when it is realized that the concept of carrying capacity is composite. It encompasses several elements, namely, the physical, the ecological, the cultural, the tourist social, and the host social carrying capacities.

Physical carrying capacity is the level beyond which the available space cannot provide for tourist visits without a clear deterioration of the tourist experience.

Ecological carrying capacity is the level of visitation beyond which unacceptable ecological impacts will occur either from the tourists or the amenities they require.

Cultural carrying capacity represents the number of visitors beyond which the cultural, historical and archaeological resources start to deteriorate in time.

Tourist social carrying capacity is the level beyond which visitor satisfaction declines unacceptably because of overcrowding.

Host social carrying capacity is the level beyond which growth will be unacceptable in terms of detriment to the host community in its traditions, ethics, value system or quality of life (WTO 1994: 23–5, 60–2).

It is important, however, to note that the carrying capacity principle is not always an absolute principle. It may expand to accommodate more tourist traffic through planning and management techniques as well as technological factors. This may be particularly true in beach tourism. While capacities vary according to the type of beach resort, availability of space on the beach may be increased in capacity by zoning, rostering or reclaiming more beach from the sea without hampering the natural balance. However, if the beach is rich in underwater corals, or if water undercurrents are active, it would be prudent not to interfere with it at all (Atherton 1993).

Carrying capacity varies according to the tourist season and is affected by tourists' behavioural patterns, the dynamic character of the environment and the changing attitudes of the host community. It is, moreover, affected by tourism policy determinants.

The question of whether an expansion of physical carrying capacity would be detrimental or not for the destination, its amenities or for the tourists themselves, needs thorough investigation and research to ascertain the pros and cons of increasing the capacity versus the physical, social and economic cost involved.

Needless to say, achieving successful tourism growth and environmental considerations cannot be separated. Tourism has to be environmentally sustainable – in both the natural and cultural environments – in order to be economically and socially sustainable.

Basic guidelines for achieving sustainable tourism

Generally speaking, sustainable tourism requires a broad vision which encompasses larger time- and space-frames than what would be traditionally required in normal tourism planning and decision making. It is not sufficient to implement such general tourism planning principles as, for example, zoning, clustering, integration between attractions and facilities, interdependence between attractions and facilities, interdependence between natural and cultural attractions, multiple ways to provide accessibility, elasticity, diversity and complementarity, cost-benefit analysis, and applications of carrying capacity.

In addition, certain guidelines have to be followed to achieve sustainable tourism.

- A general tourism policy incorporating sustainable tourism objectives at the national, regional and local levels should be followed.
- Parameters established for the planning, development and operation of tourism should be cross-sectional and integrated, involving various government departments, public and private sector companies, community groups and experts, thus providing the widest possible safeguards for success.
- In planning tourism development projects, primary consideration should be given to the protection of natural and cultural assets with due regard to the appropriate socioeconomic uses and human impacts on the physical and built environments.
- An assurance should be forthcoming that all tourism participants will follow ethical and sound behavioural and conservative rules regarding nature, culture, economy, community value systems, political patterns, social groupings and leadership.
- The distribution of tourism development projects, while abiding by a national tourism development strategy, should be undertaken with equity in mind, spreading fairly the benefits of tourism among various regions and areas including those depressed.
- Public awareness of the benefits of tourism and how to mitigate its negative impacts should be pursued.
- Local people should be encouraged to assume leadership roles in planning and development, with the necessary assistance from government, business, financial institutions and universities.
- Throughout all stages of tourism planning, development and operation, a scientific and applied assessment, monitoring and mediation programme should be carefully followed to make it possible for all those benefiting from tourism development projects to respond to changes in markets and tourism products; this is necessary if sustainability of tourism development projects is to be achieved.

Needless to say, sustainable tourism should be understood as a form of economic development that is geared to:

- ameliorating the quality of life of the host community particularly in areas in need

- assuring a better quality of tourist services
- guaranteeing a higher level of tourism development that is compatible with the environment and fulfils the aspirations of the citizens economically, culturally, socially, psychologically and politically

AGENDA 21 AND THE TRAVEL AND TOURISM INDUSTRY

Agenda 21 is a comprehensive programme of action adopted by 182 governments at the 1992 United Nations Conference on Environment and Development (UNCED), known as the Earth Summit. It provides a blueprint for securing the sustainable future of the planet into the twenty-first century and is the first document of its kind to achieve widespread international agreement and commitment to work harmoniously towards the conservation of the earth's natural resources.

The travel and tourism industry is considered the world's largest industry, generating in 1995 an estimated US$3.4 trillion in gross output, creating employment for more than 211 million people, producing 10.9 per cent of world gross domestic product (GDP), investing $693.9 billion in new facilities and equipment, and contributing more than $637 billion to global tax revenue. By the year 2005 it is estimated that the industry will have expanded its global role, generating $7.2 trillion in gross output, creating employment for 305 million people, producing 11.4 per cent of world GDP, investing $1,613 billion in new facilities and equipment, and contributing more than $1,369 billion in tax revenue (World Travel and Tourism Council 1995).

With this privileged position in the world economy, the travel and tourism industry should be recognized as a world leader in contributing to sustainable development. However, Agenda 21 did not mention travel and tourism except in a few sections. Chapter 11 (Part II, 27) for example, prescribes that governments 'promote and support the management of wildlife [and] . . . ecotourism'; Chapter 17 (Part II, 134) states that 'coastal states should explore the scope for expanding recreation and tourist activities based on marine living resources'; Chapter 36 (Part IV, 24) advocates that countries should 'promote, as appropriate, environmentally sound leisure and tourism activities . . . making suitable use of museums, heritage sites, zoos, botanical gardens, national parks, and other protected areas' (WTO 1995).

Strict adherence to the promotion of nature-oriented, specialized, humanistic and low-capacity tourism to bring about environmental quality improvement, while sound in itself, cannot live up to the promise of realizing the industry's huge expectations for worldwide improved expansion. The real benefits lie in making all forms of travel and tourism sustainable. This is the real challenge facing all travel and tourism policy makers and industry professionals.

Some key objectives to be achieved in the next ten years are as follows:

- Create socioeconomic value for primitive resources including natural beauty, wildlife, wilderness and forests as well as historical and cultural heritage, built heritage and traditional ways of life.
- Offer sufficient incentives and provide means for environmental enhancement of urban areas, especially in industrial cities.
- Provide necessary infrastructure in remote areas with tourism potential, to encourage human settlements and other economic activities to grow in harmony with tourism.
- Research and develop environmentally compatible technology that could lead to socioeconomic growth and culturally oriented advancement.
- Use international communication networks to spread messages and practices of sustainable development.
- Provide examples to other industries of how to conduct business practices that actively contribute toward sustainable development.

POLICY CONSIDERATIONS FOR TOURISM SUSTAINABILITY

Rationale

While a growing trend in policy making in many countries is to leave tourism to private enterprise, and current economic conventional thinking supports the role of market mechanisms, the pendulum shows signs of change in theory and in application. As Butler (1990: 3) puts it, 'the nature of tourism to some degree determines the nature and pattern of growth and, unless checked and controlled, will inevitably create a set of problems'. The free play of the market may lead to overreaching capacity limits and hence a lack of sustainability. Therefore, some authors stress the fact that more attention be paid to the imperfections of the market and to those specifically necessary interventions that can correct the tendencies to disequilibrium and monopolistic positions in both private and public sectors (De Kadt 1992). An example of this would be overdevelopment of the tourist supply in touristically most-frequented areas, such as Hurghada on the Red Sea in Egypt, which resulted in a cut-throat price war that has driven down prices of tourist services to a minimum.

Governments have long exercised controls in some countries and areas, most particularly in exchange rates and hotel services price levels. Governments can influence the development of tourism in certain areas by setting the conditions of investments and the scope of financial concessions, and regulating conditions of access to land, for example by allowing only long-term leases or rights of usufruct on land for foreign investors. Some governments still play an important role in providing infrastructure for tourism development, and also include tourism in their development strategies and land use plans.

State involvement in the orientation, regulation and reasonable control of the tourism industry might be deemed necessary to attain a growth rate which is regarded by governments as desirable and sometimes necessary. Such a growth rate would be higher than the market-driven rate.

Balanced tourism growth and sustainability

If past, present and future are connected by a time continuum, change, as previously mentioned, may separate these time dimensions, creating phases of different actions, reactions and outcomes. However, these would be inter-related by the forces of cause and effect. Therefore sustainability and balanced tourism growth could be deemed as two intimately related concepts. Selective growth and striking a balance between tourist supply and demand, between attractions, facilities and services, between quantity and quality, between number of arrivals and natural, social and cultural carrying capacities, are some safeguards against mismanagement of tourism. These safeguards, among other policy decisions, would lead to sustainability and avoidance or mitigation of various detriments to tourism (Wahab 1991).

It is noteworthy that implementation of Agenda 21 requires a major shift in priorities, involving a full integration of sustainable development consid-erations into economic and social policies, and a major redeployment of human and financial resources at national and international levels. In many cases there will be significant benefits from this shift; governments will spend less on healthcare and face lower clean-up costs; businesses will expe-rience cost saving from more efficient resource use; tourism destinations will enjoy long-term success; and finally our children will be guaranteed a sustainable future (WTO 1995).

This joint document translates the Agenda into an action programme for the travel and tourism industry which aims at helping national tourist administrations (NTAs), representative trade organizations and travel and tourism companies fulfil their potential to achieve sustainable development through tourism. Travel and tourism can transform the following key objec-tives into reality in the medium- and long term (WTO 1995):

- Create economic value for resources where utilization and exploitation would, in principle, yield no returns. These are natural and cultural resources such as forests, wilderness, wildlife, archaeological sites, cultural heritage and modern built heritage.
- Provide good reasons and means for environmental conservation of areas which are in need of such environmental awareness. These are mainly large cities centres and old industrial sites.
- Create employment opportunities in various regions and localities suit-able for tourism development.
- Extend necessary tourist infrastructure and superstructure to remote and depressed areas, helping to stimulate economic activities in those areas.

- Use modern communications media to transfer experiences and practices of sustainable development between nations, particularly from developed to developing countries.
- Provide examples to other industries in the adoption of business practices that contribute to sustainable development and lead ideas on sound environmental management.

Mass tourism, with the detriments it may inflict on the environment, has been severely criticized as a major environmental predator. It is therefore necessary that tourism adopts a different perspective that should be compatible, for all practical purposes, with the environment and the community in which it is active. Environmental and societal systems cannot realistically exist in an operational vacuum, but mostly coexist with judicious economic development, and address relevant needs of the host community in the same global system. Many environmentalists, however, seem to ignore or be unaware of this precondition (Farrell 1992). This is another issue that confirms the necessity to subscribe to an integrated systemic approach that resides in well conceived policy which should be properly implemented in order to guarantee sustainability.

A policy checklist for sustainable tourism development

A necessary national dialogue on sustainable tourism would result in multifarious policy directives to guide appropriate action within the context of respective mandates and capabilities. These would not only prescribe the actions to be taken by NTAs which are the most relevant, but also roles to be played by tourism industry associations, accommodation managers, tour operators and other tourist services suppliers. However, in the present context it is sufficient to deal with policy determinants that relate to the government as a whole, while acknowledging other tourism industry players' share of policy.

Governments should:

1 Conduct or cause to develop a mission statement and goals which reflect a commitment to sustainable tourism; and incorporate sustainable tourism goals with policies, decision systems, planning processes, and programmes of national tourist administrations.
2 Create and/or endorse mechanisms for effective communication, coordination and integrated resource management with other resource and environment ministries; be proactive in building tourism into sustainable development strategies, and ensuring tourism resources and values are fully identified and provided for in planning and allocation purposes.
3 Establish an integrated system of legislation, regulation, and conservation

conducive to the protection and promotion of the environmental resource base, while allowing appropriate use levels, particularly in key natural and cultural resources.

4 Coordinate efforts with regional authorities in order to avoid overlap and/or duplication of policy making in sustainable tourism issues; local jurisdiction in matters pertaining to the environment and socioeconomic aspects to be fully harmonized with the national goals and directives.

5 Mount a monitoring and evaluation programme to measure progress towards the pre-established goals.

6 Conduct or cause to be conducted national, regional and sectoral applied research on overall tourist activities and impacts including generating markets studies and socioeconomic models.

7 Assist regional and local authorities to develop their own tourism development strategies in association with environmental protection plans and community tourism awareness and involvement programmes.

8 Develop and apply sectoral and regional environmental accounting systems for tourism.

9 Implement public consultation techniques and processes in order to involve all those concerned in tourism-related decisions at the public and private sector levels.

10 Supervise the implementation of design and construction standards which ensure that tourism development projects are harmonious with physical environments and local culture.

11 Strictly apply the limitations of carrying capacities when awarding permits to tourism development projects in order to ensure sustainable levels of development.

12 Provide safeguards to ensure that tourism activities will cause no damage to the country's archaeological wealth or heritage sites.

13 Enforce strict regulations to prohibit illegal trade in historic and authentic objects, and to prevent erosion of aesthetic values and desecration of sacred sites.

14 Include tourism in national, regional and local land use planning.

15 Establish tourism advisory boards that involve responsible representatives of all tourism-related activities.

16 Require an objective and comprehensive study of environmental impacts for each tourist development project before licensing.

17 Ensure that all government departments having any role in tourism are briefed on the concept of sustainable development.

18 Implement, wherever possible, the use of new and renewable energy in tourism development projects to mitigate pollution effects. Encourage also recycling and minimization of residues in resorts.

19 Inspired by the philosophy of Total Quality Management, apply its rules

to guarantee high efficiency in administering the travel and tourism sector.

20 Give prime priority to the controlled development of environmentally and culturally important and sensitive areas.

21 Include clear provisions stressing a policy of sustainable tourism development in national, regional and local tourism development agreements.

22 Adopt and implement codes of conduct conducive to sustainability by the various tourism stakeholders.

(These requirements have been inspired by the Tourism Stream Action Strategy presented at the Globe 90 Conference in Vancouver, Canada [Nelson 1993] and by the World Conference on Sustainable Tourism, June 1995.)

SUMMARY

Tourism sustainability is a byproduct of a multitude of factors that contribute to the successful present integration and future continuity of tourism at the macro- and micro-levels in destinations. As all socioeconomic, cultural, political and environmental factors are subject to change in time and space, sustainability is therefore a relative term and not an absolute fact.

Carrying capacity is a central principle in environmental protection and sustainable tourism development. It determines the maximum use of any place without causing negative effects on resources, community, economy and culture, or reducing visitor satisfaction. Carrying capacity varies according to the tourist season and is affected by tourist behavioural patterns, the dynamic character of the environment and the changing attitudes of the host community. It is, moreover, affected by tourism policy determinants.

Whereas a growing trend in policy making in many countries is to leave tourism to private enterprise, and current economic conventional thinking supports the role of market mechanisms, the pendulum shows signs of change in theory and in application. Some developed countries such as Canada and the United States have reduced or abolished the role of the public sector in national tourism administration in favour of private enterprise. State tourism bodies remain as promotion agencies at the macro-level. This approach still represents an exception to the rule.

Needless to say, confronting the contradiction between the maximization of tourist numbers and the minimization of undesirable environmental impacts is a matter of public policy that cannot be left entirely to the discretion of the private sector. As responsibility for environmental protection ultimately lies with local planning authorities, sustainability directives should therefore be incorporated by these public authorities into their policies for tourism growth.

REFERENCES

Atherton, T. (1993) 'Measuring sustainable tourism development: problems and achievements' in *Round Table on Planning for Sustainable Tourism Development*, World Tourism Organisation, General Assembly, Bali, Indonesia.

Butler, R. W. (1993) 'Tourism: an evolutionary perspective', in *Tourism and Sustainable Development*, Waterloo, Ontario: University of Waterloo.

De Kadt, E. (1992) 'Making the alternative sustainable' in *Tourism Alternatives*, Pennsylvania: University of Pennsylvania Press, 47–75.

Edgell, D. (1993) *World Tourism at the Millennium*, Washington D.C.: US Department of Commerce.

Everyman's Dictionary of Economics (1969) London: Dent.

Farrell, B. (1992) 'Tourism as an element in sustainable development', in *Tourism Alternatives*, Pennsylvania: University of Pennsylvania Press, 115.

Jhingan, M. L. (1978) *The Economics of Development and Planning*, Delhi: Vikas Publishing House.

Nelson, J. G. (1993) 'An introduction to tourism and sustainable development', in *Tourism and Sustainable Development*, Waterloo, Ontario: University of Waterloo.

Schumpeter, E. (1942) Theory of economic development referred to in M. L. Jhingan (1978) *The Economics of Development and Planning*, Delhi: Vikas Publishing House.

Serageldin I. (1993) 'Making development sustainable', *Finance and Development*, Dec.: 6–10.

United Nations (1992) *Agenda 21*, Rio Declaration, Rio de Janeiro, Brazil.

Wahab, S. (1975) *Tourism Management*, London: Tourism International Press.

——(1988) *Planning of Tourism Resources*, in Arabic, Cairo: Hellali Press.

Wall, G. (1991) 'Tourism and sustainable development', paper presented at the Annual Meeting of the Canadian Association of Geographers, Kingston, Ontario.

World Tourism Organisation (1994) *Planning for Sustainable Development: National and Regional Tourism Planning, Methodologies and Case Studies*, Madrid: WTO.

——(1995) *Agenda 21 For the Travel and Tourism Industry*, Madrid: WTO.

World Tourism and Travel Council (1995) *New Economic Perspective*, Brussels: WTTC.

NAME INDEX

SUBJECT INDEX